Sources and Characteristics of Organic Matter in the Clackamas River, Oregon, Related to the Formation of Disinfection By-Products in Treated Drinking Water

By Kurt D. Carpenter, Tamara E.C. Kraus, Jami H. Goldman, John Franco Saraceno, Bryan D. Downing, Brian A. Bergamaschi, U.S. Geological Survey; and Gordon McGhee and Tracy Triplett, Clackamas River Water

Prepared in cooperation with Clackamas River Water and the City of Lake Oswego

Scientific Investigations Report 2013–5001

U.S. Department of the Interior
U.S. Geological Survey

U.S. Department of the Interior
KEN SALAZAR, Secretary

U.S. Geological Survey
Marcia K. McNutt, Director

U.S. Geological Survey, Reston, Virginia: 2013

For more information on the USGS—the Federal source for science about the Earth, its natural and living resources, natural hazards, and the environment, visit http://www.usgs.gov or call 1–888–ASK–USGS.

For an overview of USGS information products, including maps, imagery, and publications, visit http://www.usgs.gov/pubprod

To order this and other USGS information products, visit http://store.usgs.gov

Suggested citation:
Carpenter, K.D., Kraus, T.E.C., Goldman, J.H., Saraceno, J.F., Downing, B.D., McGhee, Gordon, and Triplett, Tracy, 2013, Sources and characteristics of organic matter in the Clackamas River, Oregon, related to the formation of disinfection by-products in treated drinking water: U.S. Geological Survey Scientific Investigations Report 2013–5001, 78 p.

Contents

Contents—Continued

Figures

Figures—Continued

Figures—Continued

Tables

Conversion Factors, Datums, and Abbreviations

Conversion Factors

Inch/Pound to SI

Multiply	By	To obtain
Length		
inch (in.)	2.54	centimeter (cm)
Area		
acre	4,047	square meter (m^2)
square mile (mi^2)	259.0	hectare (ha)
square mile (mi^2)	2.590	square kilometer (km^2)
Flow rate		
cubic foot per second (ft^3/s)	0.02832	cubic meter per second (m^3/s)
million gallons per day (Mgal/d)	0.04381	cubic meter per second (m^3/s)
Mass		
grams (g)	0.035274	ounce, avoirdupois (oz)

Temperature in degrees Celsius (°C) may be converted to degrees Fahrenheit (°F) as follows:

$$°F=(1.8×°C)+32$$

Specific conductance is given in microsiemens per centimeter at 25 degrees Celsius (µS/cm at 25°C).

Concentrations of chemical constituents in water are given either in milligrams per liter (mg/L) or micrograms per liter (µg/L).

Turbidity values in water are given either in formazin nephelometric units (FNUs) or nephelometric turbidity units (NTUs).

Ultraviolet-light absorbance is measured in nanometers.

Datums

Vertical coordinate information is referenced to the North American Vertical Datum of 1988 (NAVD 88).

Horizontal coordinate information is referenced to the North American Datum of 1983 (NAD 83).

Abbreviations

Alum	Aluminum sulfate
ACH	Aluminum chlorhydrate
BQ	Benchmark quotient
CRW	Clackamas River Water
DBP	Disinfection by-product
DBPFP	Disinfection by-product formation potential
DHBA	3,5-dihydroxy-benzoic acid
DO	Dissolved oxygen
DOC	Dissolved organic carbon
DOM	Dissolved organic matter
DWTP	Drinking-water treatment plant
ex	Excitation
em	Emission
FDOM	Fluorescent dissolved organic matter
FI	Fluorescence index
HAA	Haloacetic acid
HAA5	Total of five haloacetic acids
HAAFP	Haloacetic acid formation potentials
HIX	Humic index
LED	Light-emitting diode
LO	City of Lake Oswego
MCL	Maximum contaminant level
NMDS	Nonmetric dimensional scaling ordination
NWQL	U.S. Geological Survey National Water-Quality Laboratory
PAC	Powdered activated carbon
PCA	Principal Component Analysis
QA	Quality assurance
rpm	Revolutions per minute
S	Spectral slope
S_R	Spectral slope ratio
SRM	Standard reference material
SHAAFP	Specific formation potential for haloacetic acids
STHMFP	Specific formation potential for trihalomethanes
SUVA	Specific ultraviolet absorbance
THM	Trihalomethane
THM4	Total of four trihalomethanes
THMFP	Trihalomethane formation potential
TOC	Total organic carbon
TP	Total phosphorus
TPC	Total particulate carbon
TPN	Total particulate nitrogen
USEPA	U.S. Environmental Protection Agency
USGS	U.S. Geological Survey
UVA	Ultraviolet absorbance
UVA_{254}	Ultraviolet absorbance at 254 nanometers
VOC	Volatile organic compounds

Sources and Characteristics of Organic Matter in the Clackamas River, Oregon, Related to the Formation of Disinfection By-Products in Treated Drinking Water

By Kurt D. Carpenter[1], Tamara E.C. Kraus[1], Jami H. Goldman[1], John Franco Saraceno[1], Bryan D. Downing[1], Brian A. Bergamaschi[1], Gordon McGhee[2], and Tracy Triplett[2]

Executive Summary

This study characterized the amount and quality of organic matter in the Clackamas River, Oregon, to gain an understanding of sources that contribute to the formation of chlorinated and brominated disinfection by-products (DBPs), focusing on regulated DBPs in treated drinking water from two direct-filtration treatment plants that together serve approximately 100,000 customers. The central hypothesis guiding this study was that natural organic matter leaching out of the forested watershed, in-stream growth of benthic algae, and phytoplankton blooms in the reservoirs contribute different and varying proportions of organic carbon to the river. Differences in the amount and composition of carbon derived from each source affects the types and concentrations of DBP precursors entering the treatment plants and, as a result, yield varying DBP concentrations and species in finished water. The two classes of DBPs analyzed in this study—trihalomethanes (THMs) and haloacetic acids (HAAs)—form from precursors within the dissolved and particulate pools of organic matter present in source water.

The five principal objectives of the study were to (1) describe the seasonal quantity and character of organic matter in the Clackamas River; (2) relate the amount and composition of organic matter to the formation of DBPs; (3) evaluate sources of DBP precursors in the watershed; (4) assess the use of optical measurements, including in-situ fluorescence, for estimating dissolved organic carbon (DOC) concentrations and DBP formation; and (5) assess the removal of DBP precursors during treatment by conducting treatability "jar-test" experiments at one of the treatment plants.

Data collection consisted of (1) monthly sampling of source and finished water at two drinking-water treatment plants; (2) event-based sampling in the mainstem, tributaries, and North Fork Reservoir; and (3) in-situ continuous monitoring of fluorescent dissolved organic matter (FDOM), turbidity, chlorophyll-a, and other constituents to continuously track source-water conditions in near real-time.

Treatability tests were conducted during the four event-based surveys to determine the effectiveness of coagulant and powdered activated carbon (PAC) on the removal of DBP precursors. Sample analyses included DOC, total particulate carbon (TPC), total and dissolved nutrients, absorbance and fluorescence spectroscopy, and, for regulated DBPs, concentrations of THMs and HAAs in finished water and laboratory-based THM and HAA formation potentials (THMFP and HAAFP, respectively) for source water and selected locations throughout the watershed.

The results of this study may not be typical given the record and near record amounts of precipitation that occurred during spring that produced streamflow much higher than average in 2010–11. Although there were algal blooms, lower concentrations of chlorophyll-a were observed in the water column during the study period compared to historical data.

Concentrations of DBPs in finished (treated) water averaged 0.024 milligrams per liter (mg/L) for THMs and 0.022 mg/L for HAAs; maximum values were about 0.040 mg/L for both classes of DBPs. Although DBP concentrations were somewhat higher within the distribution system, none of the samples collected for this study or for the quarterly compliance monitoring by the water utilities exceeded levels permissible under existing U.S. Environmental Protection Agency (USEPA) regulations: 0.080 mg/L for THMs and 0.060 mg/L for HAAs.

DOC concentrations were generally low in the Clackamas River, typically about 1.0–1.5 mg/L. Concentrations in the mainstem occasionally increased to nearly 2.5 mg/L during storms; DOC concentrations in tributaries were sometimes much higher (up to 7.8 mg/L). The continuous in-situ FDOM measurements indicated sharp rises in DOC concentrations in the mainstem following rainfall events; concentrations were relatively stable during summer base flow. Even though the first autumn storm mobilized appreciable quantities of carbon, higher concentrations of DBPs in finished water were observed 3-weeks later, after the ground was saturated from additional rainfall.

---

[1]U.S. Geological Survey.

[2]Clackamas River Water.

The majority of the DOC in the lower Clackamas River appears to originate from the upper basin, suggesting terrestrial carbon was commonly the dominant source. Lower-basin tributaries typically contained the highest concentrations of DOC and DBP precursors and contributed substantially to the overall loads in the mainstem during storms. During low-flow periods, tributaries were not major sources of DOC or DBP precursors to the Clackamas River.

Although the dissolved fraction of organic carbon contributed the majority of DBP precursors, at times the particulate fraction (inorganic sediment and organic particles including detritus and algal material) contributed a substantial fraction of DBP precursors. Considering just the main-stem sites, on average, 10 percent of THMFP and 32 percent of HAAFP were attributed to particulate carbon. This finding suggests water-treatment methods that remove particles prior to chlorination would reduce finished-water DBP concentrations to some degree.

Overall, concentrations of THM and HAA precursors were closely linked to DOC concentrations; laboratory DBP formation potentials (DBPFPs) clearly showed that THMFP and HAAFP were greatest in the downstream tributaries that contained elevated carbon concentrations. However, carbon-normalized "specific" formation potentials for THMs and HAAs (STHMFP and SHAAFP, respectively) revealed changes in carbon character over time that affected the two types of DBP classes differently. HAA precursors were elevated in waters containing aromatic-rich soil-derived material arising from forested areas. In contrast, THM precursors were associated with carbon having a lower aromatic content; highest STHMFP occurred in autumn 2011 in the mainstem from North Fork Reservoir downstream to LO DWTP. This pattern suggests the potential for a link between THM precursors and algal-derived carbon. The highest STHMFP value was measured within North Fork Reservoir, indicating reservoir derived carbon may be important for this class of DBPs. Weak correlations between STHMFP and SHAAFP emphasize that precursor sources for these types of DBPs may be different. This highlights not only that different locations within the watershed produce carbon with different reactivity (specific DBPFP), but also that different management approaches for each class of DBP precursors could be required for control.

Treatability tests conducted on source water during four basin-wide surveys demonstrated that an average of about 40 percent of DOC can be removed by coagulation. While the decrease in THMFP following coagulation was similar to DOC, the decrease in HAAFP was much greater (approximately 70 percent), indicating coagulation is particularly effective at removing HAA precursors—likely because of the aromatic nature of the carbon associated with HAA precursors.

Several findings from this study have direct implications for managing drinking-water resources and for providing useful information that may help improve treatment-plant operations. For example, the use of in-situ fluorometers that measure FDOM provided an excellent proxy for DOC concentration in this system and revealed short-term, rapid changes in DOC concentration during storm events. In addition, the strong correlation between FDOM values measured in-situ and HAA5 concentrations in finished water may permit estimation of continuous HAA concentrations, as was done here. As part of this study, multiple in-situ FDOM sensors were deployed continuously and in real-time to characterize the composition of dissolved organic matter. Although the initial results were promising, additional research and engineering developments will be needed to demonstrate the full utility of these sensors for this purpose.

In conclusion, although DBPFPs were strongly correlated to DOC concentration, some DBPs formed from particulate carbon, including terrestrial leaf material and algal material such as planktonic species of blue-green algae and sloughed filaments, stalks, and cells of benthic algae. Different precursor sources in the watershed were evident from the data, suggesting specific actions may be available to address some of these sources. In-situ measurements of FDOM proved to be an excellent proxy for DOC concentration as well as HAA formation during treatment, which suggests further development and refinement of these sensors have the potential to provide real-time information about complex watershed processes to operators at the drinking-water treatment plants.

Follow-up studies could examine the relative roles that terrestrial and algal sources have on the DBP precursor pool to better understand how watershed-management activities may be affecting the transport of these compounds to Clackamas River drinking-water intakes. Given the low concentrations of algae in the water column during this study, additional surveys during more typical river conditions could provide a more complete understanding of how algae contribute DBP precursors. Further development of FDOM-sensor technology can improve our understanding of carbon dynamics in the river and how concentrations may be trending over time.

This study was conducted in collaboration with Clackamas River Water and the City of Lake Oswego water utilities. Other research partners included Oregon Health and Science University in Hillsboro, Oregon, Alexin Laboratory in Tigard, Oregon, U.S. Geological Survey National Research Program Laboratory in Denver, Colorado, and the U.S. Geological Survey Water Science Centers in Portland, Oregon, and Sacramento, California. This project was supported with funding from Clackamas River Water, City of Lake Oswego, the U.S. Geological Survey, and the Water Research Foundation.

Introduction

The Clackamas River in northwestern Oregon (fig. 1) is a valued resource to the region, supporting runs of wild steelhead and salmon and providing drinking water for nearly 400,000 people. From its headwaters near Olallie Butte south of Mount Hood, the Cascade River descends from the High Cascades flowing northwest for 82 mi to reach its confluence with the Willamette River southeast of Portland. Although 72 percent of the 940-mi^2 watershed is contained within the Mount Hood National Forest (Metro Regional Services, 1997), the lower third of the watershed drains private forests and agricultural, urban, and light industrial land that variously contribute sediments, nutrients, pesticides, and other organic compounds to the Clackamas River (Carpenter, 2003; Carpenter and others, 2008; Carpenter and McGhee, 2009). Although the river, for the most part, is exceptionally clear, it sometimes becomes turbid with sediment and organic matter from storm runoff that degrades the quality of source water at the drinking-water intakes in the lower river.

Two of the four drinking-water treatment plants (DWTPs) in the lower river—the Clackamas River Water (CRW) and the City of Lake Oswego (LO) DWTPs—use direct filtration as a means to clarify raw source water (fig. 2). Together, these two plants serve about 100,000 people. Both water utilities use chlorination simultaneously with coagulation as part of the water-treatment process. The use of chlorine as a disinfectant, although essential for pathogen control, leads to the halogenation of organic matter present in source water (Croué and others, 1999). Halogenated (chlorine- and bromine-containing) compounds form from dissolved and particulate organic carbon during water treatment and are collectively referred to as disinfection by-products (DBPs). Although only a small fraction of the organic carbon present in source water reacts to form DBPs, several DBPs have been identified as mutagenic and carcinogenic (Krasner and others, 2006; Richardson and others, 2007). For this reason, the USEPA currently regulates two classes of DBPs commonly found in drinking water—trihalomethanes (THMs) and haloacetic acids (HAAs) (U.S. Environmental Protection Agency, 2009). In addition to being a source of DBPs, organic carbon contributes to biofouling, increases chlorine demand, and can affect aesthetic qualities of water such as taste, odor, and color (Cooke and Kennedy, 2001).

Water managers are concerned about DBPs in drinking water and are interested in identifying the types of organic carbon that contribute DBP precursors in source water to better understand the potential for future deterioration in river quality resulting from a wide array of possible sources (fig. 3). Understanding the timing, sources, and composition of organic matter entering drinking-water intakes will help drinking-water utilities develop source-water-protection programs, facilitate successful and cost-effective treatment strategies, and help plan for future upgrades to treatment plants (Kraus and others, 2010).

Although much of the carbon in the watershed is contained in its forests and soils, the Clackamas River, along much of its length, is gravel bedded and provides appreciable habitat for benthic algae (periphyton) to attach and grow (see photographs 1a-e). This material contains organic carbon that may decompose to yield DBP precursors (Jack and others, 2002; Huang and others, 2009; Kraus and others, 2011). Periphyton growths occur in the Clackamas River during periods when nutrients, light, and flow conditions are favorable. At times, periphyton biomass reaches nuisance levels along river margins and causes supersaturated concentrations of dissolved oxygen (DO), high (alkaline) pH, and large daily fluctuations in pH and DO. The pH in the lower river is particularly high, and regularly exceeds the State of Oregon water-quality standard during parts of the growing season, particularly in spring (U.S. Geological Survey, 2012). Periphyton biomass is typically higher in the lower river downstream from Estacada, but nuisance biomass levels also occur in the upper river, upstream from North Fork Reservoir (fig. 1). A previous U.S. Geological Survey (USGS) study (Carpenter, 2003) found nuisance levels of benthic algae in the main-stem Clackamas River during the summer, and continuous monitoring of DO and pH—definitive algal-photosynthesis indicators—shows this problem has continued on and off for at least the past decade.

When periphyton detaches from the riverbed and becomes entrained in the flow during algal "sloughing" events, the cells and stalks of diatoms, filaments of green algae, and colonies of blue-green algae (Cyanobacteria) in varying states of decomposition enrich the river with organic carbon. This algae has the capacity to clog water intakes and negatively affect drinking-water quality through the production of tastes and odors and algal toxins (Graham and others, 2010). Algae also contribute carbon that contains DBP precursors (Graham and others, 1998; Jack and others, 2002; Kraus and others, 2011). Phytoplankton (floating algae) also occasionally form blooms during summer in the two primary reservoirs, Timothy Lake in the headwaters of the Oak Grove Fork and North Fork Reservoir on the mainstem (see photographs 2a-c), which can also contribute DBP precursors (Kraus and others, 2011).

Decomposition products of terrestrial plant material, including leaves from deciduous trees, conifer needles, and other plant material contained in soils are certainly a source of organic carbon in the Clackamas River. High rainfall leads to saturated conditions and flow of water through organic-rich surface soils. High rainfall combined with steep topography in much of the basin causes erosion, and landslides are particularly common during large storms. The epic February 1996 rain-on-snow event, for example, produced over 200 landslides in the Fish Creek Basin alone (DeRoo and others, 1998), and much of that material deposited in the main-stem Clackamas River near the upstream end of North Fork Reservoir. Decomposing vegetation buried in that debris could be another source of DBP precursors, along with wastewater from three municipal treatment plants, septic-tank effluents, and other potential sources.

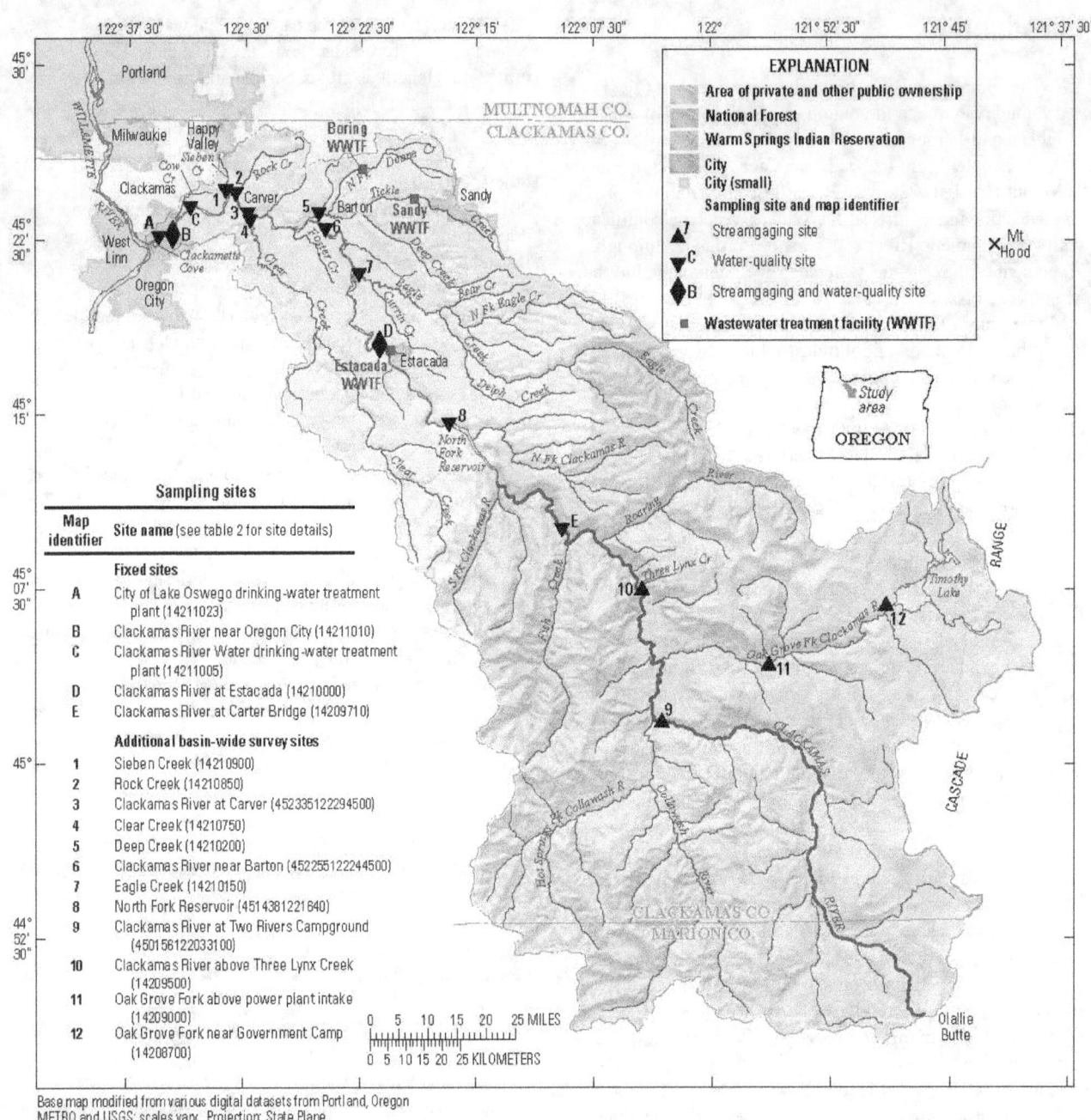

Figure 1.   Land ownership of the Clackamas River basin, Oregon, and the sampling-site locations.

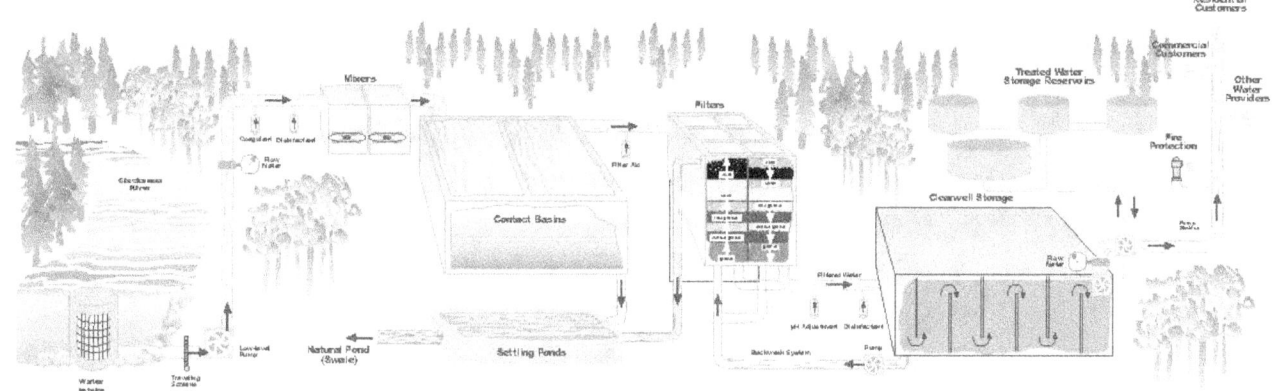

Diagram courtesy of Clackamas River Water.
http://crwater.com/images/stories/2011WQReport_web_version_2.pdf

Figure 2. The direct-filtration process at the Clackamas River Water drinking-water treatment plant, Clackamas, Oregon.

Photograph 1a. Riffles support many types of benthic algae in the upper Clackamas River upstream of Carter Bridge. (Photograph by Kurt Carpenter, U.S. Geological Survey, June 2010.)

Photograph 1b. *Prasiola* sp. (green algae) on the large cobbles in the Clackamas River downstream of River Mill Dam near Estacada. (Photograph by Kurt Carpenter, U.S. Geological Survey, April 23, 2006.)

Photograph 1c. *Prasiola* sp. (green algae) on a cobble from the Clackamas River downstream of River Mill Dam near Estacada. (Photograph by Kurt Carpenter, U.S. Geological Survey, June 2010.)

Photograph 1d. Stalked diatoms (*Cymbella* sp.) on a cobble from the Clackamas River downstream of River Mill Dam near Estacada. (Photograph by Kurt Carpenter, U.S. Geological Survey, June 2010.)

Photograph 1e. *Nostoc* sp. (blue-green algae) colony balls on a cobble from the upper Clackamas River upstream of Carter Bridge. (Photograph by Kurt Carpenter, U.S. Geological Survey, June 2010.)

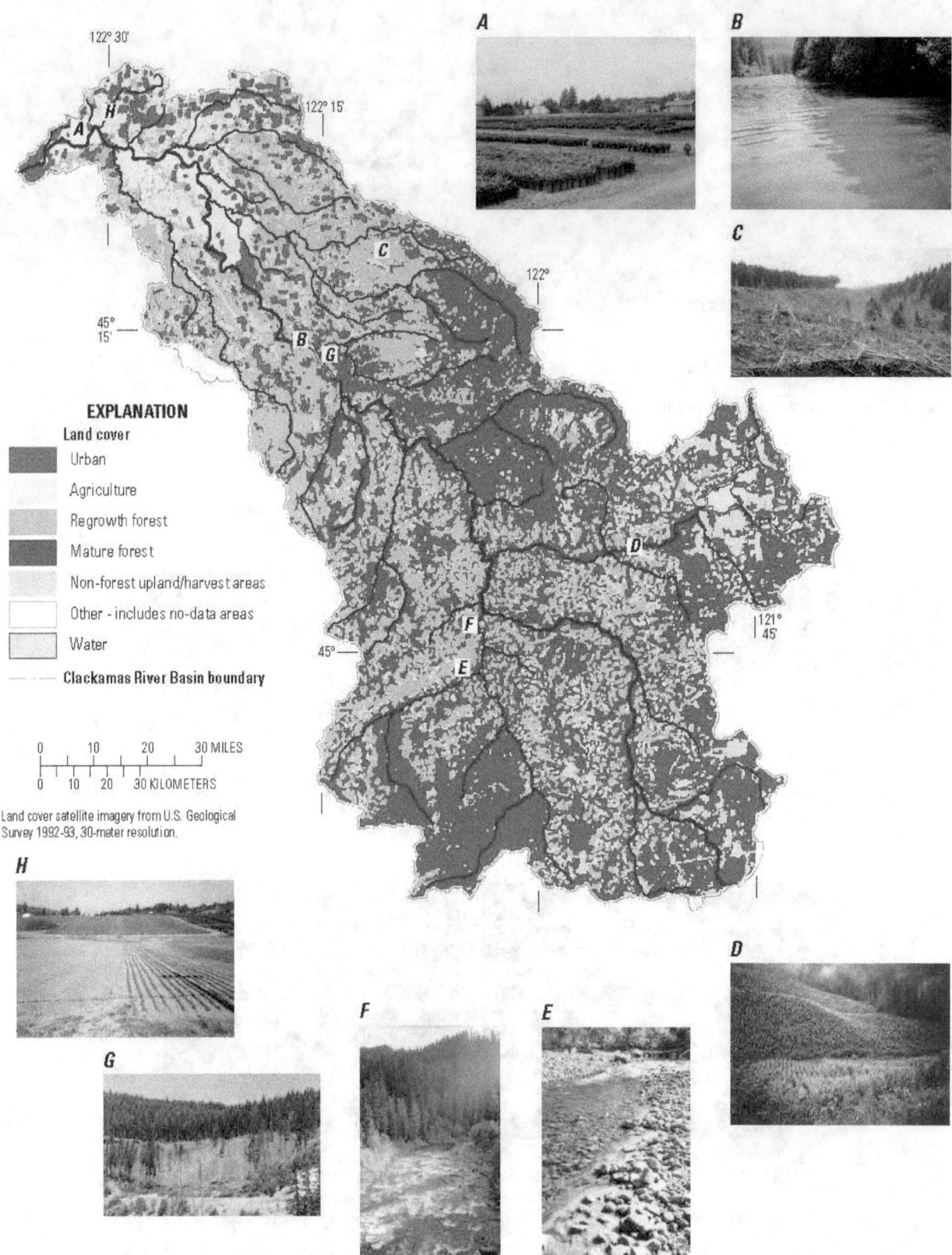

Figure 3.   Land cover and potential sources of organic carbon that contribute disinfection by-product precursors to the Clackamas River, Oregon. (Photographs show (*A*) irrigated container nursery in the Deep Creek basin, (*B*) blue-green algae bloom in North Fork Reservoir, (*C*) timber harvest unit in the Eagle Creek basin, (*D*) regrowth forest in the Oak Grove Fork basin, (*E*) filamentous green algae in the Collawash River, (*F*) riparian forest in the Collawash River basin, (*G*) timber harvest unit adjacent to North Fork Reservoir, and (*H*) row crops in the Rock Creek basin.)

Photograph 2a-c. Blue-green algae (*Anabaena flos-aquae*) blooms in North Fork Reservoir and Timothy Lake. (Photographs by Kurt Carpenter, U.S. Geological Survey, August 31, 2010.)

Photograph 2b.

Photograph 2c.

Disinfection By-Products and Historical Trends in Disinfection By-Product Concentrations

DBPs are known to be carcinogenic and are indicated to cause increased risks of reproductive and developmental problems in humans (Richardson and others, 2007). For these reasons, the USEPA regulates the total concentration of four THMs (THM4) and five HAAs (HAA5) in finished (treated) drinking water (U.S. Environmental Protection Agency, 2006). In 2004–05, as part of the USGS National Water-Quality Assessment Program, source and finished water from the CRW DWTP were sampled for a variety of organic compounds, including THMs. Compounds including pesticides and gasoline hydrocarbons were commonly detected in finished water; relative to regulatory standards, however, THMs, were the compound class of most concern (Carpenter and McGhee, 2009). Even though THM4 concentrations were always below current regulatory thresholds, THM4 concentrations may be higher when organic carbon is elevated in source water such as during periods of active rainfall runoff, algal blooms, or periphyton sloughing events—which were not targeted for sampling during that study.

Compliance monitoring data from the CRW and LO DWTPs show THMs in finished drinking water have increased to some degree over the past 20 years (fig. 4). The cause of this increase is not known, and data to evaluate potential causes, such as trends in historic total organic carbon (TOC) concentrations, are not available. Historically, the two highest THM4 concentrations were measured in mid-to-late June 1997 and 2000. It is possible these high THM4 concentrations were caused by sloughed benthic algae because this is a time of year when sloughing events have occurred (as in 2004 and 2005) (U.S. Geological Survey, 2012).

To better understand this increase in DBP concentrations, basin-specific information is needed regarding the sources of carbon that form the DBPs and factors that contribute to carbon losses from watersheds so that management strategies can be developed and successfully targeted. Information on the type of carbon present can provide insights into the chemical reactions that take place during chlorination that might shape water-treatment strategies to control DBPs. This could be especially important if the types of carbon that contribute to DBPs increase in source water or if the USEPA regulations become more restrictive.

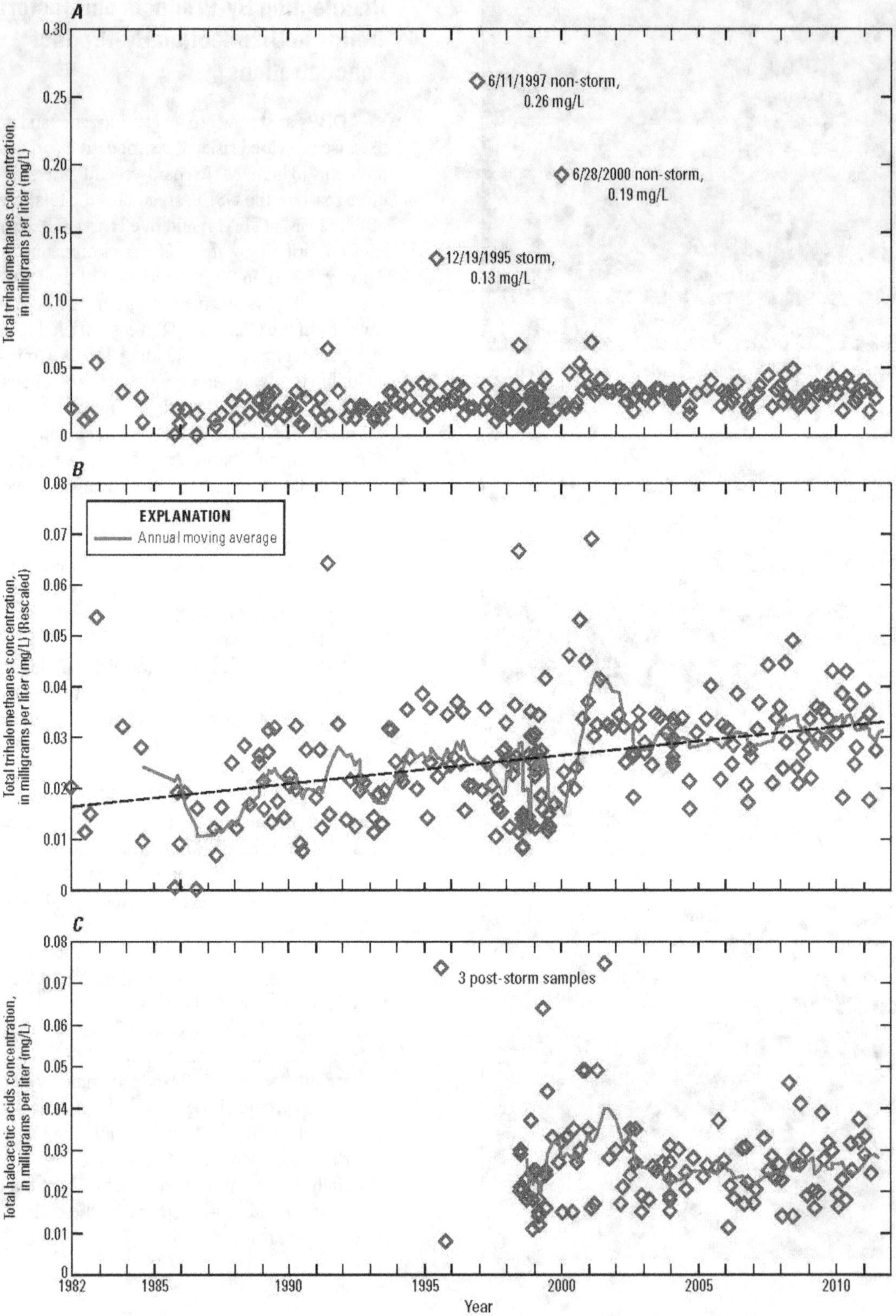

A

6/11/1997 non-storm,
0.26 mg/L

6/28/2000 non-storm,
0.19 mg/L

12/19/1995 storm,
0.13 mg/L

B

EXPLANATION
—— Annual moving average

C

3 post-storm samples

Figure 4.   Historical trends (1982–2011) in disinfection by-product concentrations in finished drinking water from the Clackamas River Water and City of Lake Oswego direct-filtration treatment plants, Clackamas River basin, Oregon. (*A*) Total trihalomethanes (THM4). (*B*) Total trihalomethanes (THM4) with y-axis rescaled. (*C*) Total haloacetic acids (HAA5).

Sources of Organic Matter and Disinfection By-Product Precursors

The DBP precursor pool is a subset of the bulk organic-matter pool present in source water. Possible sources of organic matter in the Clackamas River basin are shown in figure 3. Previous studies have shown terrestrial plants and soils (allochthonous sources) and algae and macrophytes (autochthonous sources) contribute organic matter and DBP precursors to surface waters (Aiken and Cotsaris, 1995; Reckhow and others, 2004). As described previously, algae are a possible source of organic carbon in the Clackamas River. While less is known about the propensity for periphyton to form DPBs, phytoplankton is a well-known source of dissolved organic carbon (DOC) and DBP precursors, especially HAAs (Jack and others, 2002; Nguyen and others, 2005; Kraus and others, 2011).

The amount and reactivity of organic matter entering a DWTP is a function of the amount and composition of material entering the water throughout the watershed, as well as environmental processes such as biodegradation, photodegradation, sedimentation, and sorption that may take place during transport through the river system. The types of DBPs that form are controlled by the physiochemical properties of the carbon molecules and the complex reactions that occur with disinfectants such as chlorine, along with coagulation treatment, pH, temperature, bromide concentration, and other factors (Crepeau and others, 2004).

The amount of organic matter in a water sample is typically determined by measuring carbon concentration, assuming that half the organic matter pool is made up of carbon. TOC is commonly characterized by laboratory measurements of whole (unfiltered) water; however, this method has a tendency to under-report carbon concentrations (Aiken and others, 2002). In this report, TOC was derived by summing laboratory measurements of DOC and total particulate carbon (TPC). In addition to these concentration-based constituents, the composition of dissolved organic matter (DOM) was characterized using absorbance and fluorescence spectrophotometry, and continuous in-situ fluorescence was used as an indicator of DOC concentration.

Use of Optical Properties to Characterize Dissolved Organic Matter

Spectral optical property measurements such as absorbance and fluorescence can be used to determine the amount of DOC in water and to broadly characterize dissolved organic matter (DOM) composition (Hudson and others, 2007; Fellman and others, 2010; Matilainen and others, 2011). Shifts in the spectral response of optical properties can help identify sources of carbon within watersheds and inform watershed management (Kraus and others, 2010; Beggs and others, 2011; Bridgeman and others, 2011). Absorbance measures the amount of light absorbed by material in a water sample at specified excitation (ex) wavelengths, and fluorescence measures the light that is re-emitted (emission (em) wavelengths). Depending on the type of material present, the spectral properties can be diagnostic of certain types of organic matter. If material derived from different sources provides a unique "signature", important sources of carbon within a watershed can be identified. Studies have demonstrated that continuous, in-situ fluorescence measurements can be used as a reliable surrogate for DOC concentration, and advances are ongoing to improve understanding of how changes in carbon composition affect freshwater systems using this approach (Bergamaschi and others, 2005, 2012; Spencer and others, 2007; Saraceno and others, 2009; Pellerin and others, 2012).

Absorbance and Fluorescence Spectroscopy

The measurement of ultraviolet absorbance at 254 nm (UVA_{254}) has been used by the drinking-water industry as a proxy for DOC concentration for several decades (Edzwald and others, 1985; Rathbun, 1996; Korshin and others, 1997; Sadiq and Rodriguez, 2004). In addition to providing information about DOC concentration, absorbance data can provide insight into the chemical make-up of the DOM pool (table 1). For example, UVA_{254} normalized by DOC concentration, also known as "specific" UVA (SUVA, reported in units of liters per milligram-meters, L/mg-m), has been correlated with DOM aromatic content (Weishaar and others, 2003). Similarly, the spectral slope of the absorbance curve has been shown to relate to aromatic content and molecular weight. For example, decreasing spectral slope between 275 and 295 nm is associated with higher aromatic content and increasing molecular weight. Spectral slope has also been shown to change upon irradiation (Helms and others, 2008; Spencer and others, 2009).

As with absorbance, the fluorescence response and intensity at a single ex/em wavelength pair can be related to DOC concentration; the presence of different peaks, peak slopes, changes in the ratios of ex/em pairs, carbon normalized values, or shifts in peak maxima have been shown to provide information about DOM character and origin (Coble, 2007; Hudson and others, 2007; Stedmon and Bro, 2008). The fluorescence index (FI), calculated as the ratio of em 470 to 520 nm and ex 370 nm, has been widely used to indicate relative contributions of terrestrial- and algal-derived DOM (table 1).

Table 1. Optical properties commonly used as indicators of carbon composition.

[**Abbreviations:** DOC, dissolved organic cargon; DOM, dissolved organic matter; L, liters; mg, milligrams; m, meters; nm, nanometers; ex, excitation; em, emission; –, not applicable]

Parameter	Description	Calculated	Units
Specific Ultraviolet Absorbance (SUVA)[1]	Positively correlated with aromatic carbon content.	Absorbance at 254 nm normalized to DOC concentration	(L/mg-m)[1]
Spectral Slope (S_{X-Y})[2,3,4,5]	Higher values indicate low molecular weight DOM and (or) decreasing aromaticity; lower values indicate DOM with a higher molecular weight and (or) higher aromatic content.	Nonlinear fit (exponential function) of the absorption spectrum over a specified wavelength range from X to Y nm.	nm[-1]
Spectral Slope Ratio (S_R)[4,6]	Positively correlated to DOM molecular weight; generally increases upon irradiation.	Ratio of spectral slopes: $S_{275-295}$ divided by $S_{350-400}$	–
Fluorescence Index (FI)[7,8]	Higher values associated with microbial sources such as extracellular release and leachate from bacteria and algae; lower values associated with terrestrially-derived soil and plant organic matter, typical values range from 1.3–1.9.	The ratio of em wavelengths at 470 nm to 520 nm, obtained at ex 370 nm	–
Humification Index (HIX)[9,10,11]	Indicator of humic substance content or extent of humification; based on the idea that the emission spectra of fluorescing molecules will shift toward longer wavelengths due to lower H:C ratios as humification of DOM proceeds.	Area under the em spectra 435–480 nm divided by the peak area 300–345 nm, at ex 254 nm	–

[1]Weishaar and others (2003).
[2]Blough and Del Vecchio (2002).
[3]Obernosterer and Benner (2004).
[4]Helms and others (2008).
[5]Twardowski and others (2004).
[6]Spencer and others (2009).
[7]McKnight and others (2001).
[8]Cory and McKnight (2005).
[9]Zsolnay and others (1999).
[10]Ohno (2002).
[11]Fellman and others (2010).

FI values obtained in the laboratory typically range from about 1.3 to 1.9; lower FI values are associated with terrestrial soil and plant organic matter—highly processed material having greater aromatic content and higher molecular weight—while higher FI values are associated with lower molecular weights and lower aromatic content indicative of algal and microbial sources (McKnight and others, 2001; Cory and others, 2010). Qualitative information can also be derived from the identification of fluorescence regions that have been linked to different DOM pools such as humic and fulvic acids, protein like substances, and phytoplankton-derived material (Stedmon and others, 2003; Coble, 2007; Hudson and others, 2007).

In-Situ Fluorometers as Proxies for Dissolved Organic Carbon and Disinfection By-Product Precursor Concentrations

The use of fluorescence around ex 370/em 460 nm has been shown to have similar, possibly better, predictive ability for DOC concentration compared to absorbance measurements (Nakajima and others, 2002; Coble, 2007; Kraus and others, 2010). The continuous measurement of fluorescing DOM (FDOM) with in-situ fluorometers has been successfully used to provide a high-resolution proxy for DOC concentration (Downing and others, 2009; Saraceno and others, 2009; Pellerin and others, 2012). However, because only a subset of the DOM pool fluoresces and this fluorescing pool is a function of DOM composition, the relation between DOC concentration and FDOM needs to be validated for each watershed over the complete range of riverine conditions. The effects of optical density, DOC concentration, temperature, and turbidity on FDOM also need to be evaluated and accounted for (Lakowicz, 2006; Downing and others, 2012).

Because DBP precursors are a sub-set of the bulk carbon pool, FDOM may also serve as a good proxy for THM and HAA precursor concentrations and, thus, for finished-water THM and HAA concentrations. However, the composition of the DOM pool can affect this relation because it affects the fraction of the DOC pool that reacts to form DBPs and the fraction of the DOM pool that fluoresces. Currently, the information regarding the relation between FDOM and DBP formation is limited (but see Nakajima and others, 2002; Hua and others, 2007, 2010; Beggs and others, 2009; Marhaba and others, 2009; Kraus and others, 2010). Given that the optically-active aromatic fraction of DOM typically dominates the DBP precursor pool in terrestrial environments, there is good reason to believe that in most cases there is a strong relation between these constituents. Furthermore, data from the McKenzie River in Oregon, a similar Cascade Mountain

drainage, showed FDOM was a better predictor of THMFP and HAAFP than DOC concentration, suggesting a strong overlap between DOM moieties that are fluorescent and react with chlorine to form DBPs (Kraus and others, 2010).

To date, commercially available fluorometers intended to assess DOM dynamics are centered near ex 370/em 460 nm, referred to as "Peak C," a humic region of the excitation-emission (fluorescence) matrix (EEM) commonly identified in surface waters. While measurement of a single ex/em pair provides information about DOC concentration, information about the composition of the DOM pool is currently only available using bench-top fluorometers; sample scans using these instruments produce nearly 2,300 ex/em pairs, which are depicted in the EEM diagrams (fig. 5).

Ongoing developments in light-emitting diode (LED) manufacturing technology have resulted in the availability of light sources with lower excitation wavelengths into the deep ultraviolet spectrum. Pairing the nearly monochromatic output of these LEDs with a wide array of optical filters makes novel development of miniaturized in-situ fluorometers with different excitation/emission pairs possible. Instruments designed using narrow band-pass filters allow for more focused emission spectra and, thus, also a more specific emission–fluorescence signal around a narrower excitation/emission region. Sensors designed to measure different regions of EEMs can provide signal ratios that may indicate carbon composition changes due to varying proportions of fluorescence associated with different pools of organic matter (humic peaks compared to amino-acid-like peaks, for example). When deployed continuously and in real-time, these sensors present new opportunities to gain insights about how rivers function and identify what factors affect water quality and source-water supplies, especially when a multitude of constituents are measured simultaneously.

Recent work in an agricultural watershed demonstrated the in-situ FDOM sensor accurately predicted DOC concentrations throughout a precipitation and watershed runoff event, where DOC concentration cycled from a baseline of 2 mg/L to a peak value of 10 mg/L and back again to base-flow levels (Saraceno and others, 2009). Several recent studies from a range of environments including wetlands, tidal marshes, and forested watersheds produced predictive relations between in-situ FDOM values from a WET Labs™ colored DOM (CDOM) fluorometer and laboratory-determined DOC concentrations (Downing and others, 2008, 2009; Bergamaschi and others, 2012; Pellerin and others, 2012). This Clackamas study is, to our knowledge, the first to deploy a continuously-operated multi-channel in-situ FDOM sensor within a drinking-water treatment plant intake with simultaneous periodic measurement of DBPs in finished water.

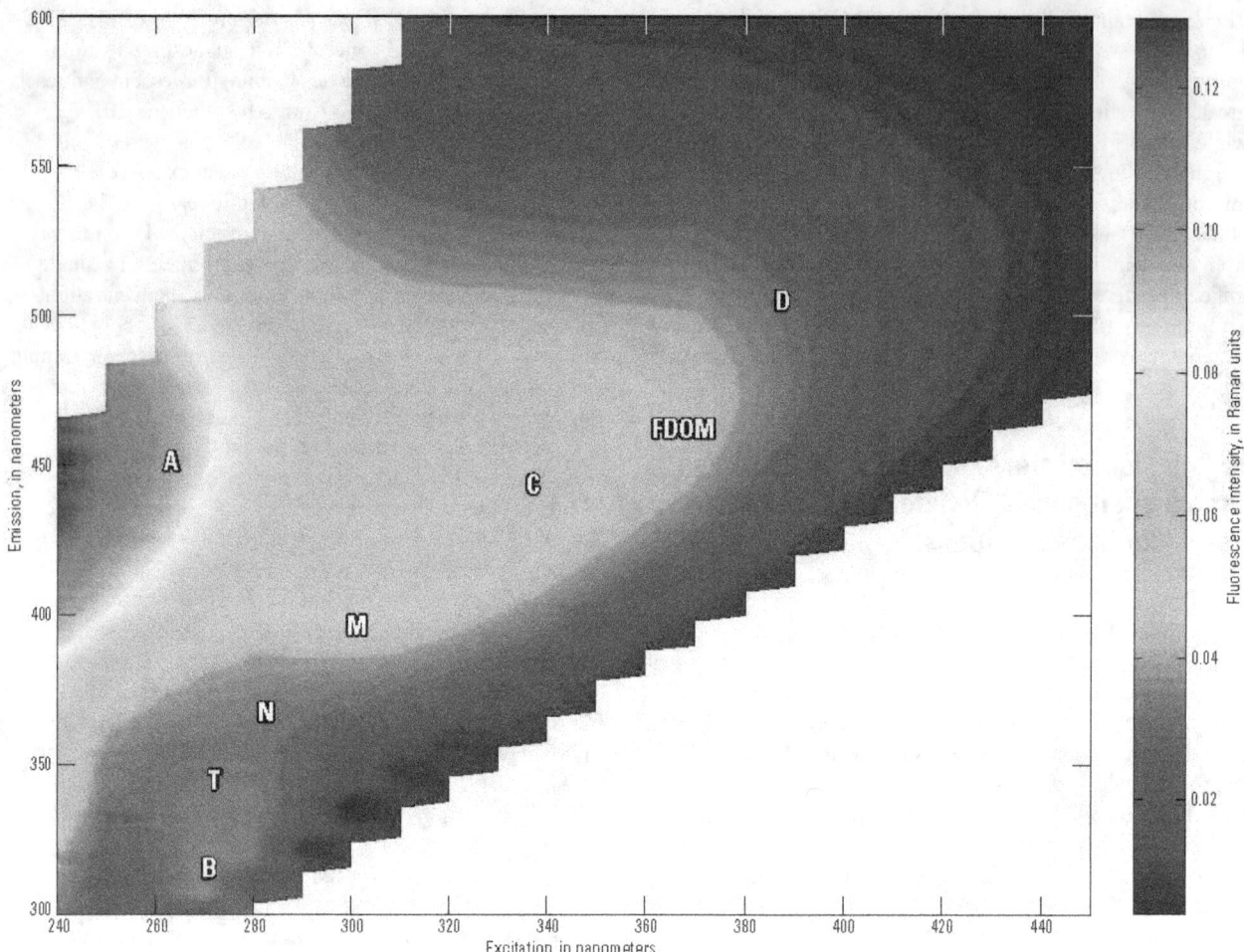

Figure 5. Example excitation–emission matrix showing the general locations of selected fluorescence peaks. (The contour plot is made up of 2,291 unique excitation-emission (ex/em) pairs. Individual letters indicate the location of common peaks associated with different pools of dissolved organic matter, which are described in figure 21. FDOM indicates the region measured with standard in-situ sensors. Example shown is from a source-water sample collected from the Clackamas River Water DWTP on April 4, 2010.)

Study Objectives and Approach

The five main study objectives were to (1) characterize the seasonal quantity and quality of organic carbon in the Clackamas River and its primary tributaries; (2) relate the amount and composition of organic carbon to the formation of DBPs at sites throughout the watershed and in finished drinking water; (3) evaluate the major suspected sources of DBP precursors in the watershed, including tributaries, North Fork Reservoir, litter/soils, and algae; (4) assess the use of optical properties, including in-situ FDOM, for estimating DOC and DBP precursor concentrations; and (5) assess the treatability of DOC and DBP precursors by conducting "jar-test" experiments at one of the drinking-water treatment plants.

The approach consisted of (1) continuous in-situ fluorescence monitoring of source water at the CRW DWTP in the lower river from April 2010 to September 2011; (2) monthly water sampling at four main-stem sites (table 2) for DOC, TPC and total particulate nitrogen (TPN), total and dissolved nutrients, and optical properties including absorbance and fluorescence (table 3); (3) four basin-wide surveys: spring high flow, summer base-flow period, late-summer reservoir drawdown, and the first "initial" autumn storm (table 4); and (4) four treatability tests on source water using standard jar-tests procedures to evaluate the potential for coagulant and powdered activated carbon (PAC) to reduce DOC and DBP precursor concentrations. During each basin-wide survey, DBP formation potentials (DBPFP) for THMs (THMFPs) and HAAs (HAAFPs) were measured to determine DBP precursor concentrations in surface water in the major lower-basin tributaries, North Fork Reservoir at the log boom, and the Clackamas River mainstem.

Table 2. Description of continuous monitors and sites sampled during monthly sampling and basin-wide surveys, Clackamas River basin, Oregon.

[Sampling sites–short name: Site locations are shown in figure 1. Abbreviations: NWIS, National Water Information System; CRW, Clackamas River Water; LO, City of Lake Oswego; DWTP, drinking-water treatment plant; USFS, U.S. Forest Service; ~, approximately; –, not applicable]

Sampling activity				Sampling sites— short name	River mile	NWIS site No.	NWIS site name	Dominant land cover
Monthly	Basin-wide surveys	Continuous water-quality monitor	Streamflow gage					
				Main-stem Clackamas River / CRW / LO DWTP sites				
			X	Clackamas River at Two Rivers Campground		450156122033100	Upper Clackamas River at Two Rivers C.G., OR	Forested (fir-hemlock dominant); USFS land
			X	Clackamas River agove Three Lynx Creek	47.8	14209500	Clackamas River above Three Lynx Creek, OR	Forested (fir-hemlock dominant); USFS land
X		X		Clackamas River at Carter Bridge	40.8	14209710	Clackamas River at Carter Bridge, near Estacada, OR	Forested (fir-hemlock dominant); USFS land
X	X	X	X	Clackamas River at Estacada	23.1	14210000	Clackamas River at Estacada, OR	Mix of forest, rural residential, and urban
	X			Clackamas River at Barton	13.4	452255122244500	Clackamas River near Barton (QW site), OR	Mix of forest, rural residential, and urban
	X			Clackamas River at Carver	7.9	452335122294500	Clackamas River at Carver, OR	Mix of forest, rural residential, and urban
X	X	X		CRW DWTP–Source water	3.1	14211005	Clackamas River at Clackamas, OR (source)	Mix of forest, rural residential, urban, and industrial
X	X			CRW DWTP–Finished water	–	14211006	Clackamas River at Clackamas, OR (finished)	Mix of forest, rural residential, urban, and industrial
X		X	X	Clackamas River near Oregon City	1.6	14211010	Clackamas River near Oregon City, OR	Mix of forest, rural residential, urban, and industrial
X	X			LO DWTP–Source water	0.9	14211023	Clackamas River at Gladstone, OR (source)	Mix of forest, rural residential, urban, and industrial
X	X			LO DWTP–Finished water	–	14211024	Clackamas River at West Linn, OR (finished)	Mix of forest, rural residential, urban, and industrial

Table 2. Description of continuous monitors and sites sampled during monthly sampling and basin-wide surveys, Clackamas River basin, Oregon, 2010–11.—Continued

[Sampling sites–short name: Site locations are shown in figure 1. Abbreviations: NWIS, National Water Information System; CRW, Clackamas River Water; LO, City of Lake Oswego; DWTP, drinking-water treatment plant; USFS, U.S. Forest Service; ~, approximately; –, not applicable]

Sampling activity				Sampling sites— short name	River mile	NWIS site No.	NWIS site name	Dominant land cover
Monthly	Basin-wide surveys	Continuous water-quality monitor	Streamflow gage					
Tributary sites								
			X	Oak Grove Fork near Government Camp	15.5	14208700	Oak Grove Fork near Government Camp, OR	Forested (fir-hemlock dominant); USFS land
			X	Oak Grove Fork above powerplant intake	6.1	14209000	Oak Grove Fork above powerplant intake, OR	Forested (fir-hemlock dominant); USFS land
	X			Eagle Creek	16.7	14210150	Eagle Creek at Bonnie Lure State Park, OR	Forested (alder-fir dominant), regrowth forests/Christmas trees, rural residential
	X			Deep Creek	12.1	452340122251000	Deep Creek at Hwy 224, OR	Agricultural (nurseries, Christmas trees, berries, other crops), regrowth forests
	X			Clear Creek	8	14210750	Clear Creek at Carver, OR	Forested (alder-fir dominant) and rural residential, Christmas trees; regrowth forests
	X			Rock Creek	6.4	14210850	Rock Creek near Carver, OR	Agricultural (nurseries, berries), rural residential
	X			Sieben Creek	5.8	14210900	Sieben Creek at Hwy 224, OR	Residential, industrial
North Fork Reservoir sites								
X				1 - Surface (0.5 foot depth)	31.2	451438122164001	North Fork Res at Boom Station 1 (Depth 1), OR	Forested (fir-hemlock dominant)
X				2 - Metalimnion (mid-depth)	31.2	451438122164002	North Fork Res at Boom Station 1 (Depth 2), OR	Forested (fir-hemlock dominant)
X				3 - Release depth (~80 feet)	31.2	451438122164003	North Fork Res at Boom Station 1 (Depth 3), OR	Forested (fir-hemlock dominant)
X				4 - Hypolimnion (off bottom)	31.2	451438122164004	North Fork Res at Boom Station 1 (Depth 4), OR	Forested (fir-hemlock dominant)

Table 3. Field and laboratory data collected during monthly samplings and basin-wide surveys, Clackamas River basin, Oregon, 2010–11.

[Sampling site locations are shown in figure 1. Constituents defined in table 6. **Abbreviations:** TPN, total particulate nitrogen; TPC, total particulate carbon; chl-a, chlorophyll-a; CRW, Clackamas River Water; DOC, dissolved organic carbon; DWTP; LO, City of Lake Oswego DWTP; DWTP, drinking-water treatment plant; DBP, disinfection by-product; FP, formation potential; RM, river mile; ~, approximately]

Monthly	Basin-wide surveys	Targeted DBP sources[1]	Sampling sites	Field parameters	Total nutrients	TPN/TPC	DOC	Dissolved nutrients	Optical properties[2]	Water-column chl-a	Benthic chl-a	DBPFP unfiltered[3]	DBPFP filtered[3]	DBPs[4]
			Main-stem Clackamas River / DWTP sites[5]											
X	X		Clackamas River at Carter Bridge	X	X	X	X	X	X	X	X	X	X	
X	X		Clackamas River at Estacada	X	X	X	X	X	X	X	X	X	X	
	X		Clackamas River at Barton Bridge	X	X	X	X	X	X	X	X	X	X	
	X		Clackamas River at Carver Bridge	X	X	X	X	X	X	X	X	X	X	
X	X		CRW DWTP–Source water	X	X	X	X	X	X	X		X	X	
X	X		CRW DWTP–Finished water				X		X					X
X	X		LO DWTP–Source water	X	X	X	X		X					
X	X		LO DWTP–Finished water				X		X					X
			Tributary sites											
	X		Eagle Creek	X	X	X	X	X	X	X	X	X	X	
	X		Deep Creek	X	X	X	X	X	X	X	X	X	X	
	X		Clear Creek	X	X	X	X	X	X	X	X	X	X	
	X		Rock Creek	X	X	X	X	X	X	X	X	X	X	
	X		Sieben Creek	X	X	X	X	X	X	X	X	X	X	
			North Fork Reservoir sites											
	X		1 - Surface (0.5 foot depth)	X	X	X	X	X	X	X		X	X	
	X		2 - Mid-depth (mid-depth)	X	X	X	X	X	X	X		X	X	
	X		3 - Release depth (~80 feet)	X	X	X	X	X	X	X		X	X	
	X		4 - Near bottom (off bottom)	X	X	X	X	X	X	X		X	X	
			Upper Clackamas River/Forest											
		X	Clackamas River upstream from Carter Bridge		X	X	X		X	X			X	
		X	Alder Flat Trail		X	X	X		X	X			X	

[1] Includes four types of benthic algae (periphyton) and water extracts from two types of forest leaf litter/soils (Douglas fir and Red alder).

[2] Optical properties include absorbance and fluorescence.

[3] DBPFPs include determination of THMs and HAAs.

[4] DBPs, DOC, and optical properties were analyzed in finished water during monthly samplings and basin-wide surveys.

[5] U.S. Geological Survey operates three continuous water-quality monitors in the Clackamas River (fig. 1), including water temperature, specific conductance, dissolved oxygen, pH, turbidity (all stations), and water-column chlorophyll-a (at Estacada [RM 23.1] and Oregon City [RM 1.6] only). Data from the Oregon City station were used to indicate quality of source water for both DWTPs.

Table 4. Summary of data collection including continuous dissolved organic matter fluorescence deployments and discrete water-sample collection in the Clackamas River basin, Oregon, 2010–11.

[**Sampling date(s):** Range of dates is shown in parentheses. **Abbreviation:** FDOM, fluorescing dissolved organic matter]

Sampling date(s)	FDOM sensor		Sampling activity			Flow condition	Description
	WET Labs™ WETStar	Turner Designs™ Cyclops	Monthly	Basin-wide surveys	Reservoir sampling		
04-14-2010	X		X			Medium high	Springtime melt-runoff
05-(11–12)-2010	X		X	X		Medium high	
06-05-2010	X		X			Very high	
07-08-2010	X		X			Medium low	
08-03-2010	X		X		X	Low	Algal bloom in North Fork Reservoir
09-(03–07)-2010	X		X		X	Low-medium low	
10-10-2010	X		X	X		Medium low	Initial autumn storm
11-01-2010	X		X			Medium	Major flush storm event
12-07-2010	X		X			Medium	
01-18-2011	X		X			Very high	
02-16-2011	X		X			Medium high	
03-16-2011	X	X	X			High	
04-19-2011	X	X	X			Medium high	
05-24-2011	X	X	X			Medium high	
06-29-2011	X	X	X			Medium	Periphyton sloughing
08-02-2011	X	X	X			Medium low	Algal bloom in North Fork Reservoir
09-(08–09)-2011	X	X	X	X	X	Medium low	Summer base flow
09-(16–22)-2011		X		X	X	Medium low	Timothy Lake drawdown
October 2011		X				Medium	
November 2011		X				Medium	
December 2011		X				Medium low	

Study Site Descriptions

The study sites (tables 2 and 3) included 7 in the mainstem: three sites in the upper basin including North Fork Reservoir at the log boom, and 5 sites in the lower basin, including two DWTPs that obtain source water from the river in the lower 3 miles (fig. 1). Five tributaries that drain a range in land cover, from mostly forested to mostly urban, also were sampled. There are two major reservoirs in the basin—Timothy Lake in the headwaters of the Oak Grove Fork and North Fork Reservoir on the middle mainstem. Timothy Lake is typically maintained at full pool during most of summer, and then drawn down in September to accommodate autumn and winter precipitation and to augment flows during the low-flow period. While no sampling occurred at Timothy Lake, North Fork Reservoir was sampled each year during summer blooms of blue-green algae. In contrast to Timothy Lake, North Fork is operated as a run-of-the-river reservoir, along with two other downstream diversions and dams that serve Faraday and River Mill hydroelectric facilities. Residence times are thus typically short for North Fork (on average, hours up to approximately 7 days, possibly longer depending on temperature stratification) compared to Timothy Lake (on average, approximately 8.5 months). More information on these reservoirs and the associated hydroelectric project is presented in Carpenter (2003).

Discrete Sampling

Discrete water samples were collected from within the watershed and at the DWTPs during each monthly and all basin-wide and reservoir samplings (table 3). Seven additional discrete water samples were collected at the CRW DWTP intake over a range in flow/turbidity conditions and analyzed for DOC concentration, absorbance, and fluorescence to compare laboratory measurements to the in-situ FDOM sensor response.

Watershed Samples

Water samples were collected from wadeable stream sites (table 2) using the Equal-Width Increment method, where the entire stream cross section was sampled using a depth-integrating sampler (Edwards and Glysson, 1999). Water samples were collected from unwadable main-stem sites using a D-74 sampler in mid-stream (at bridge sites). Samples from non-wadable, bridgeless sites were collected as grab samples from the main flow directly into 1-L combusted amber glass bottles. Point samples from each site were composited and homogenized in a 16-L Teflon™ churn splitter and placed on ice prior to dispensing subsamples for nutrients, DOC, optical properties, DBPFPs, and water-column chlorophyll-*a*. Samples for DOC and optical properties (absorbance and fluorescence) were filtered using 25-mm, 0.7-μm precombusted Whatman™ GF/F filters with a glass filtration unit. The filtrate was collected into combusted amber-glass bottles with Teflon™-lined caps and stored in the dark at 4°C until analyzed. All samples were analyzed within 5 days of collection.

Dissolved nutrient samples were filtered through a Pall™ 0.45 μm pore-size capsule filter. Total and dissolved nutrient samples were immediately frozen and stored at -20°C until analysis. Whole-water samples for total phosphorus and total nitrogen were obtained from the churn splitter. Samples for TPN and TPC were dispensed from the churn splitter into 125-mL baked amber-glass bottles. The contents were filtered using precombusted 25-mm Whatman™ GF/F filters (0.7-μm pore size), noting the filtrate volume. Filters were frozen and shipped to the USGS National Water-Quality Laboratory (NWQL) in Denver, Colorado, for analysis. A known volume of sample water was filtered for water-column chlorophyll-*a* using the same GF/F filters as above. Filters were wrapped in aluminum foil, frozen at -4°C, and analyzed within 30 days at the Oregon Graduate Institute in Portland, Oregon. Samples for DBPFP were collected into 1-L baked amber glass bottles and placed on ice. Formation potentials were evaluated on filtered and unfiltered samples. Filtered samples were passed through 142-mm GF/F filters (0.7 μm pore size) into 1-L baked amber glass bottles. DBPFP samples were acidified to a pH of 2.0 units using reagent-grade concentrated HCl to prevent possible microbial transformations in the samples prior to analysis. Samples were shipped on ice within 24 hours to the organic chemistry laboratory at the USGS California Water Science Center (CAWSC) in Sacramento, California, where they were refrigerated at 4°C until further processing.

Source-Water and Finished-Water Samples

Raw intake "source" water was collected from both DWTPs. At the CRW DWTP, source-water samples were collected adjacent to the FDOM sensor inside the gravity-fed vault adjacent to the river, where water enters before being pumped up to the treatment plant. Samples at the LO DWTP were collected from taps inside the treatment plant. Treated or "finished" water was collected 90 minutes after source-water sampling to account for time-of-travel through each plant.

Because DBPs commonly continue to form within the distribution system, on one occasion, as part of the regular monthly sampling, an extra DBP sample was collected from

within the distribution systems of both water utilities, at the location where historical maximum DBP concentrations have occurred. These samples were collected on November 2, 2010, 1 day following the collection of the regular finished-water samples to account for time-of-travel.

Finished-water samples for determination of DBPs were collected by CRW and LO personnel into baked amber glass bottles with a Teflon™ septa to help remove bubbles. Bottles contained a quenching agent (65 mg sodium thiosulfate for THMs, final concentration 1.5 g/L, and 150 mg ammonium chloride for HAAs, final concentration of 1.18 mg/L) to fully oxidize the residual chlorine and stop further DBP formation. THM samples were collected directly into 40-mL baked amber glass vials used for collection of volatile organic compounds, using care to slowly and completely fill each vial to prevent bubble formation. HAA samples were collected into 125-mL baked amber glass bottles. Sample collection and processing of other source- and finished-water samples were conducted as described above for watershed samples.

Benthic Algae (Periphyton)

Periphyton samples were collected for biomass (chlorophyll-*a*) and dominant species composition at four main-stem sites and five tributary sites in July to September 2010. For the tributaries, samples were collected from 10 representative rocks/cobbles in shallow riffle areas; in the mainstem, samples were collected along wadeable portions of the channel margin. All samples were collected using USGS methods (Moulton and others, 2002) described in detail in Carpenter (2003). A known area of periphyton (approximately 4-5-in. diameter circle) was scraped from the top of each rock into a plastic dishpan using a plastic bristle brush. Multiple samples were composited and transferred to plastic 1-L bottles and placed on ice until processing (described below).

Laboratory Analytical Procedures

Organic Carbon

DOC concentrations were measured in duplicate using the platinum catalyzed persulfate wet-oxidation method on an O.I. Analytical Model 700 TOC Analyzer™ (Aiken and others, 1992), which produced standard errors of ±0.2 mg/L carbon. The standard curve, consisting of a minimum of five standards over the range of interest, was repeated for every 10–12 water samples; reported values are the average of two duplicate measurements on each sample.

TPC samples were collected and analyzed using USEPA method 440.0. Particles from a known volume of water were filtered onto 25-mm GF/F filters (0.7-μm pore size), which were frozen at -4°C and shipped to the USGS NWQL for analysis. TOC was calculated as the sum of DOC and TPC.

Absorbance and Fluorescence

The absorption spectra was measured between 200 and 750 nm on filtered samples at constant 25°C temperature with a J&M TIDAS™ spectrophotometer, using a 1-cm quartz cell and distilled water for the blank. SUVA was calculated by dividing UVA_{254} by DOC concentration and is reported in L/mg-m units (Weishaar and others, 2003). Spectral slopes were calculated using a non-linear fit of an exponential function to the absorption spectrum over specified wavelength ranges (275–295 nm and 350-400 nm, for example) as described by Twardowski and others (2004). The spectral slope ratio (SR) was calculated as the ratio of $S_{275-295}$ to $S_{350-400}$ (Helms and others, 2008) (table 1). Because of the marked decrease in UVA absorbing DOM following treatment, an exponential fit could not be applied to the absorbance curve; thus, spectral slope was not calculated for finished-water samples.

Fluorescence EEMs were measured on filtered samples in a 1-cm cuvette at 20°C with a SPEX Fluoromax–4 spectrofluorometer (Horiba Jobin Yvon, Edison, New Jersey) using a 150W Xenon™ lamp, a 5-nm band pass, and 0.05-second integration time. Fluorescence intensity was measured at excitation wavelengths of 240 to 450 nm at 10-nm intervals and emission wavelengths of 300 to 600 nm at 2-nm intervals on room-temperature samples (25°C) in a 1-cm quartz cell. The resulting matrix consisted of 2,291 individual ex–em pairs that form the basis of the EEM diagrams. EEMs were blank corrected, instrument corrected, and normalized to the daily water Raman peak area, and the Rayleigh scatter lines were removed. The FI was calculated as the ratio of emissions at 470 to 520 nm at an excitation of 370 nm (McKnight and others, 2001; Cory and others, 2010). The Humic Index (HIX) was calculated by dividing the sum of fluorescence intensities at emission 436–480 nm by emission 300–346 nm at excitation 254 nm (Zsolnay and others, 1999).

Disinfection By-Product Formation Potentials

DBPFP was determined on filtered and unfiltered samples to determine the relative importance of the dissolved and particulate fractions. Both THMFP and HAAFPs were determined following a version of USEPA Methods 502.2, 510.1, and 552.2 as described by Crepeau and others (2004). Briefly, the method involved a 7-day reaction time, pH buffered at 8.3, temperature held at 25°C, and final, residual-free chlorine concentration restricted to between 2–5 mg/L. This incubation period provides information on the total DBP precursor pool and should result in concentrations of THM and HAA reflective of potential distribution-system concentrations based on residence times within the systems. For consistency, the same quenching agents used for the finished-water samples were used to quench the chlorination reaction—sodium thiosulfate for THMFP and ammonium chloride for HAAFP. However, instead of using solid material, quenching agents were dissolved in deionized water and added to the reaction vials (approximately 3 μL/mL, 0.3 percent by volume) to obtain equivalent concentrations to those used for the finished-water sample vials. Chlorine dosing and quenching were conducted at the USGS Laboratory at the CAWSC in Sacramento, California. Determination of four THMs and five HAAs was then performed by Alexin Laboratories as described below.

For quality-assurance (QA) purposes, one blank and seven samples of standard reference material (SRM) were submitted to the laboratory for analysis. As described by USEPA Method 5710B, a freshly prepared solution of anhydrous 3,5-dihydroxy-benzoic acid (DHBA, 0.039 g/L) was made to test the precision of laboratory chlorine dosing, quenching, storage, and DBP analysis. According to this method, a 7-day reaction period should result in approximately 0.119 mg/L THM as chloroform with essentially no contribution from bromide-containing THMs. With the exception of the first batch of samples sent to Alexin Laboratories, each batch of samples submitted to the lab contained at least one DHBA SRM sample.

Disinfection By-Product Concentrations in Finished Water

DBP analyses were performed on finished-water and formation-potential samples at Alexin Laboratories (ORELAP Certification ID# OR100013) in Tigard, Oregon. Analyses included four THMs (chloroform, Cl_3CH; bromoform, Br_3CH; bromodichloromethane, Cl_2BrCH; and dibromochloromethane, $ClBr_2CH$) and five HAAs (monochloroacetic acid, MCAA; dichloroacetic acid, DCAA; trichloroacetic acid, TCAA; bromoacetic acid, BrAA; and dibromoacetic acid, Br_2AA) following the same methods used for DWTP compliance monitoring—USEPA method 524.2 for THMs and method 6251B for HAAs. For this study, samples were analyzed within 2–14 days for THMs and within 5–7 days for HAAs.

In this report, the sum of individual THMs (THM4) and HAAs (HAA5) is commonly used because these metrics form the basis of the DBP drinking-water regulations. Benchmark quotients (BQs), the ratio of the concentration in finished water divided by the maximum contaminant level (MCL), were calculated to compare DBP concentrations to USEPA standards. For example, BQ values of 0.5 and 1.0 would indicate concentrations at one-half the MCL and at the MCL, respectively. Although BQs provide an indication of how close individual concentrations are to the standards, the actual standards are based on the annual running averages in quarterly sampling at the DWTP and within the distribution system.

"Specific" THMFPs and HAAFPs (STHMFPs and SHAAFPs) values were calculated for filtered and unfiltered water by dividing formation potentials by sample DOC or TOC concentration, respectively, and are reported in milligrams of DBP (THMs or HAAs) per milligram of carbon.

Some studies refer to these carbon-normalized formation potentials as DBP "yields" (Summers and others, 1996). This expression indicates the average reactivity of carbon in a water sample to form DBPs during chlorination.

Water-Column Chlorophyll-*a*

Water-column chlorophyll-*a* measurements were conducted at Oregon Health and Science University using a Turner 10-AU fluorometer according to the manufacturer's procedure. A known amount of whole water was filtered through 0.7-μm GF/F filters. The filters were stored at -20°C for no more than 30 days prior to analysis. The filters were steeped in 5-mL of 90 percent acetone for 24 hours and stored at -20°C prior to analysis. The fluorometer was zeroed with 90 percent acetone, and standard curves were generated. Standard- and regular-sample fluorescence was measured before and after addition of 3 drops of 10 percent HCL. Values were "blank corrected" by subtracting the background fluorescence of the 90 percent acetone solution. Data were entered into the manufacturer's equation to calculate chlorophyll-*a* concentrations.

Benthic Algal Chlorophyll-*a* and General Species Composition

In the laboratory, periphyton samples were homogenized in an electric blender and transferred to an 8-L churn splitter; subsamples were removed using a 5-mL pipette. Known aliquots were transferred onto 0.7-μm 45-mm GF/F filters under a mild vacuum. Each filter was folded into quarters, placed in a plastic petri dish, wrapped in aluminum foil, and frozen at -4°C until analysis. Filters were hand ground in 90 percent acetone to extract the chlorophyll-*a* pigment, and samples were analyzed using standard methods (fluorometry with acid correction) at the Oregon Water Science Center (ORWSC). A Certified chlorophyll-*a* standard solution from the USGS NWQL was used to generate standard curves to extrapolate sample concentrations. Samples also were analyzed for dominant species composition at the ORWSC using a Leica microscope and current taxonomic references.

Nutrients

Dissolved nutrient analyses for nitrate, nitrite, soluble reactive phosphate, ammonium, and silicate were performed at Oregon Health and Science University using a 5-channel 2008 model Astoria-Pacific Segmented Continuous Flow Injection Analyzer designed for spectrophotometric analysis of nutrients in freshwater. In order to attain low detection limits, the protocols followed the manufacturer's recommendations (detailed in Whitledge and others, 1986), with adjustments for low-concentration detection as outlined in Sakamoto and others (1990). All measurements were quality controlled using commercially purchased standard reference samples approved by the USGS. Dissolved organic nitrogen, total

dissolved nitrogen, total dissolved phosphorus, total nitrogen, and total phosphorus were analyzed following an alkaline persulfate digestion followed by the colorimetric procedures outlined above. TPN samples were collected and analyzed using USEPA method 440.0. Particles from a known volume of water were filtered onto 25-mm 0.7-μm GF/F filters, which were frozen at -4°C and shipped to the USGS NWQL for analysis.

Treatability Experiments Using Jar Tests

During each of the four basin-wide sampling events, jar tests were conducted on CRW DWTP source water using a Phipps and Bird Stirrer Model 7790-400 following standard protocols to assess treatability, defined here as the percentage of the DOC and DBP precursor pool removed from solution by coagulation. Water was collected from the intake vault and composited in an 18-L Teflon™ churn splitter, and subsamples were dispensed into test jars. Jar tests were designed to simulate the typical treatment at the CRW DWTP to remove suspended particles, primarily the addition of aluminum sulfate (alum) and aluminum chlorhydrate (ACH). The amount of coagulant added was adjusted as needed to an "optimum dose," defined as the amount of coagulant per liter required to reach the point of zero charge as determined by an in-line streaming current monitor. Thus, when TOC (DOC + TPC) increases, higher coagulant doses were typically applied. Because CRW chlorinates during coagulation, there is no opportunity within the drinking-water treatment train to isolate the effects of coagulation alone on DOC and DBP precursor removal, making laboratory jar tests a natural choice for helping address this question. In addition, up to 5 mg/L PAC may be added during treatment to control tastes and odors. This practice might also reduce DBPs in finished water, so PAC was included as one of the treatments.

Three treatments (two replicates each) were compared: (1) coagulants (alum and ACH) applied at optimum dose; (2) coagulants (alum and ACH) applied at optimum dose along with 5 mg/L PAC; and (3) a control that received no coagulation. The amount of coagulant required to remove the maximum amount of DOC was determined during the time of source-water collection by the in-line streaming current monitor at the plant.

The jar-test protocol loosely followed that described by Shin and others (2008). Briefly, 2-L samples were placed in each of six jars and mixed at 300 revolutions per minute (rpm) for 5 seconds. Coagulants and PAC were added, and samples were mixed at 150 rpm for 3 minutes to provide rapid mixing, followed by mixing at 25 rpm for 15 minutes to enable flocculation. Samples were allowed to settle for at least 15 minutes and then filtered through 0.7-μm GF/F filters into baked amber glass containers to generate subsamples for DOC, absorbance, fluorescence, and DBPFP. The pH of coagulated sample water was checked to confirm that coagulation did not lead to a substantial drop in pH (greater than 0.5 standard units).

On each of the four sampling dates, to verify the coagulation rates used represented the optimal dose, an additional series of six jar tests were conducted with dosing rates ranging from 25 to 200 percent of the optimal dose. Results from those tests verified the coagulant dosages used were appropriate and attained maximum DOC removal (±0.1 mg/L, data not shown). Jar-test results were compared to information obtained by comparing source-water and finished-water DOC concentrations and changes in optical properties.

Continuous Real-Time Measurement of Streamflow, Field Parameters, and Fluorescent Dissolved Organic Matter

The existing monitoring network of five USGS streamgages and three continuous water-quality monitors (fig. 1) provided flow, water temperature, specific conductance, DO, pH, and turbidity data which provided information on river conditions during the study in near real-time. In addition, at two main-stem sites—Estacada and Oregon City—water-column chlorophyll-*a* fluorescence was also measured continuously. One additional temporary site, Clackamas River near Clackamas, USGS station 14211005, was established within the CRW DWTP intake vault to monitor the quality of the source water (actual pump water) every 30 minutes for water temperature, turbidity, and FDOM.

FDOM measurements were made using two standard fluorometers: (1) a Wet Labs™ WETStar flow-through sensor deployed from April 4, 2010, to October 20, 2011, and (2) a Turner Designs™ "open-faced" Cyclops-7 sensor (see photograph 3) deployed from April 14, 2011, to January 31, 2012. In addition, three custom-built Cyclops-7 sensors were deployed along with the standard Cyclops-7 sensor (table 5).

One of these sensors was designed to detect fluorescence around ex 270/em 340 nm, a signal associated with amino acid/protein-like (peak T) fluorescence and, in the Clackamas River, expected to indicate the presence of algal-derived DOM. Two additional custom sensors with relatively narrow band-passes centered around emissions 470 and 520 nm both at excitation 370 nm were used to calculate the FI that, based on prior studies, can indicate a shift between microbial or algal-derived and terrestrial-derived sources of DOM.

All four of the Turner Designs™ Cyclops-7 fluorometers were mounted on a Cyclops 6 data logger equipped with a wiper. Water temperature, turbidity, and specific conductance also were measured at the CRW DWTP intake using a Yellow Springs Instruments, Inc., 6MS multi-probe sonde. All sensors were housed within the CRW intake vault. The sensors were raised and lowered into the water in a non-reflective black plastic basket using a pulley system. The basket was adjusted to be approximately 3 ft below the water surface, which was typically about 10-15 ft off the bottom.

The WET Labs™ WETStar FDOM sensor was deployed in an unfiltered flow-through configuration using a 12-volt submersible pump (see Saraceno and others, 2009). One drawback of this configuration is that fouling of the optics is not mitigated through the use of a wiper; therefore, optics were cleaned manually every 2-4 weeks with lens paper. Prior to each measurement, the pump flushed the sensor with approximately 2–3 L of water for 10 seconds. Then while pumping continued, the measurements were recorded for 30 seconds at 1 hertz to yield a set of burst data. The burst data were reduced to a single data point using descriptive statistics. Typical variation within a burst was less than 1 percent. The median of the last 20 samples within the burst was used in the final dataset. Unlike the WETStar™, the Turner Designs™ C6 was outfitted with a wiper that cleaned each sensor before each measurement. In addition, as with the WETStar™, optics

Table 5.   Description of in-situ fluorescing dissolved organic matter (FDOM) fluorometers sensors deployed in the Clackamas River basin, Oregon, April 2010–September 2011.

[Excitation-emission wavelengths (± bandpass values) given in nanometers as full width half maximum. **Abbreviation:** FI, Fluorescence Index]

Sensor	Excitation	Emission	Primary fluorescence peak[1]	Abbreviation
WETLabs™ WETStar Fluorometer[2]	370 (±10)	460 (±120)	Peak C	WETStar
Turner™ Cyclops-7 Fluorometer[3]	365 (±60)	470 (±30)	Peak C	Standard FDOM
Turner™ Cyclops-7 Custom Fluorometer #1[3]	370 (±20)	470 (±20)	Peak C	FI-A
Turner™ Cyclops-7 Custom Fluorometer #2[3]	370 (±20)	520 (±20)	Peak C	FI-B
Turner™ Cyclops-7 Custom Fluorometer #3[3]	270 (±12)[4]	340 (±20)	Peak T	Peak T

[1]Fluorescence peaks as commonly defined in the literature (Coble, 2007, for example).

[2]Sensor deployed April 01, 2010, to October 20, 2011.

[3]Sensor deployed April 14, 2011, to January 31, 2012.

[4]There is no optical filter on the deep ultraviolet light-emitting diode light source.

were cleaned manually every 2-4 weeks with lens paper. Data reduction was performed using burst data acquired at a rate of one sample every 11 seconds. Real-time data were recorded using a Campbell Scientific™ model CR 1000 data logger (see photograph 4) programmed for the suite of sensors deployed at the site. The data logger provided the power-distribution controls (turning sensors on and off), time stamp, and internal logging of all sensor data at predetermined sampling frequencies. The data logger was interfaced with a cellular modem (Sierra Wireless™ RAVEN XT-V) to allow remote data acquisition and troubleshooting. The data logger transmitted data to the USGS server every 2–4 hours using a Campbell Scientific™ model COM220 telephone modem. This system provided data in near real-time, which was examined using the USGS Data Grapher program (U.S. Geological Survey, 2011).

Because of the diel demand on water consumption within the service district, pumps within the intake typically turn off at night, even though the sensors continued to operate. For this reason, pump records were used to extract FDOM data only when the pumps were operating (a threshold of 7 Mgal/d or about 11 ft³/s was used to indicate when the pumps were on). This assured the measurements of FDOM within the intake vault were representative of the river and not that of possible impoundment effects in the intake of the river and not influenced by potential impoundment effects in the vault.

Water temperature and suspended particles affect fluorescence measurements (Zepp and others, 2004; Lakowicz, 2006; Saraceno and others, 2009). To correct for these effects, experimentally-derived correction factors were applied using the approach outlined in Downing and others (2012). When available, temperature and turbidity data from the co-located YSI instrument were used to make the corrections; otherwise data were obtained from the Clackamas River at Oregon City site about 1.5 mi downstream. Although temperature and turbidity values sometimes differed between these two locations, the relations were linear and reasonably well-correlated for overlapping data ($r=0.94$, $n=13,112$ and $r=0.99$, $n=15,929$, respectively). Applying temperature and turbidity corrections to raw FDOM data resulted in about an additional 8–15 percent signal recovery for baseline periods and up to an additional 23 percent signal recovery for periods characterized by high-flow storm events. High turbidity (greater than 300 NTUs) during one January storm required the most corrections that increased FDOM values by as much as 55 percent. Temperature- and turbidity-corrected FDOM data were converted to quinine sulfate dihydrate equivalents on the basis of laboratory calibration tests for quality-control purposes (Downing and others, 2012; Pellerin and others, 2012). These data were converted to DOC concentration in milligrams per liter using the near-linear relation between discrete DOC concentrations and in-situ FDOM values.

Photograph 3. The Turner Cyclops™ "C6" multi-sensor sonde deployed in the Clackamas River at the Clackamas River Water drinking-water treatment plant intake. (Photograph from Turner Designs, http://www.turnerdesigns.com/products/submersible/c6-multi-sensor.)

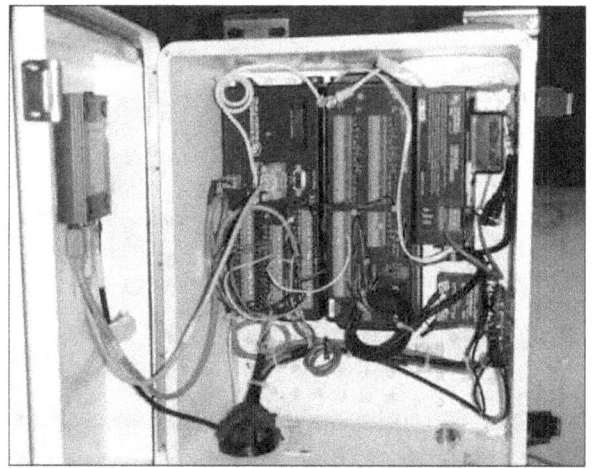

Photograph 4. The data and telecommunications system used to operate the real-time continuous instrumentation at the Clackamas River Water drinking-water treatment plant intake. (Photograph by Kurt Carpenter, U.S. Geological Survey, May 5, 2010.)

Because the effects of temperature and turbidity on the three custom sensors were not evaluated, corrections could not be applied to these data. Data from the two Peak C custom sensors were converted into quinine sulfate dihydrate equivalents using laboratory-based calibration data for each sensor, and the ratio of these sensors was calculated as Sensor FI-A/Sensor FI-B and is referred to as $FI_{in-situ}$. Data from the Cyclops-7 Peak T sensor was also not adjusted for turbidity or temperature, but data were used in terms of changes in intensity relative to other measurements such as DOC and FDOM.

Statistical Analyses

Correlations and Exploratory Data Analyses

Spearman rank correlations were used to evaluate nonparametric relations among key constituents for various site groupings (mainstem, tributaries, source/finished water) using PRIMER-E, version 6 (Clarke and Gorley, 2006). Correlations were considered significant at probabilities less than 5 percent ($p<0.05$). To better understand patterns within the data and among groups of selected sites/samples for exploratory data analyses, multivariate analyses (principal component analysis [PCA] and non-metric dimensional scaling ordination [NMDS]) were performed using PRIMER-E. For brevity, these results are not shown.

Confidence and prediction intervals for relations between concentrations of HAA5 and $FDOM_{in-situ}$ were generated using SAS (SAS Institute, Inc., 2003). The regression-line parameters (slope and intercept) for the linear relation between the $FDOM_{in-situ}$ and laboratory-derived HAA5 concentrations in finished water were applied to the time-series $FDOM_{in-situ}$ values to generate a predicted time series for HAA5 concentrations.

Carbon and Disinfection By-Product Formation Potential Loads and Yields

To quantify and better understand the carbon and DBP precursor sources during each of the basin-wide synoptic samplings, instantaneous loads were calculated by multiplying DOC, TOC, and DBPFP concentrations in milligrams per liter by streamflow in cubic feet per second, then multiplying by 2.447 to convert the units to kilograms per day. The carbon and DBPFP loads were compared to those at the CRW DWTP to evaluate the contribution from a particular site to that found in source water. The instantaneous yields (in loads per day per square kilometer) were calculated by dividing the instantaneous load by the basin area, thus providing an estimate of the loading "intensity" at each of the sampling sites at one point in time. It should be stressed that this is an exploratory analysis to help identify watershed DBP precursor sources; it is not intended to convey the magnitude of DBP formation under actual treatment conditions, and may not necessarily be representative of the full range of carbon concentrations or DBP yields at a site.

Parallel Factor Analysis (PARAFAC)

PARAFAC was used to decompose the fluorescence signatures in the corrected EEMs into unique fluorescent groups and provide more information about the character of the DOM pool (Bro, 1997). PARAFAC analysis is a type of three-way PCA that resolves absorption and emission spectra of orthogonal fluorophore groups (components) and determines loadings (proportional to concentrations) of each component. The component percentages were calculated by dividing the component loading of individual components by the sum of the component loadings to reveal qualitative differences between samples (Andersson and Bro, 2000). The Stedmon and Bro (2008) PARAFAC tutorial was used to develop the model with Matlab 2009A using the N-way toolbox, version 6.1 (Bro, 1997; Andersson and Bro, 2000). Goodness of fit was determined by visual inspection of the measured, modeled, and residual (measured minus modeled) EEM spectra, as well as by good agreement between duplicates.

PARAFAC models were validated using a combination of (1) outlier identification, (2) residual analysis, (3) component validation, and (4) replication by split-half analysis (Stedmon and Bro, 2008). The model suffered from difficulties in distinguishing weak fluorescent components from instrument noises in the low-excitation wavelengths between 240 and 250 nm, which was remedied by censoring all the EEMs at 250 nm; signals below this cut-off value were not considered. The model used a non-negative constraint to help alleviate the instrument noise and detection in samples having low fluorescence. Several model iterations were performed with different subsets of samples, mostly to examine the effects of the finished-water samples on the model results. Results indicated the finished-water samples did not contain significantly different components compared to the untreated source-water samples; thus, they were included in the model. This is consistent with a previous study that examined oxidation effects of chlorine on fluorescence (Beggs and others, 2009).

Overview of River Conditions During 2010–11

The study period spanned two exceptionally wet years with higher-than-average streamflow compared to the 103-year period of record (fig. 6). Water temperatures in the lower Clackamas River at Oregon City were about 2–3°C lower during the first half of both summers and, in 2011, about 4–6°C lower in mid-July compared with the previous 8 years (data not shown). These conditions may have limited the degree to which algae affected source-water quality during this study. Although the continuous FDOM monitor indicated sharp rises in DOC concentrations resulting from rainfall events, DOC concentrations in the mainstem were generally low, typically about 1.0–1.5 mg/L and occasionally up to about 2.5 mg/L during early-season storms. Summary statistics for selected water-quality, optical, and DBP data are presented in table 6.

The first monthly sampling was conducted in mid-April 2010 (table 4), during the last snowmelt period of the season. The first basin-wide survey was conducted in May when flows were still moderately high, about 3,500 ft³/s at Oregon City (fig. 7). The last significant storm event of the water year, in early June, produced nearly 3 in. of rain. Samples were collected near the peak in flow when streamflow was about 13,000 ft³/s at Oregon City (fig. 7). Later that summer, a blue-green algae bloom in North Fork Reservoir (*Anabaena flos-aquae* and *Microcystis aeruginosa*) was sampled in August and again in September (table 4). This bloom prompted the Oregon Health Authority to issue a human-health recreational advisory for the reservoir on September 2, 2010. The second basin-wide survey on October 10, 2010, followed the first *initial* runoff-producing rain event in autumn. About an inch of rain fell over a 2-day period that produced slightly higher turbidity (about 5 FNU) in the lower mainstem (fig. 7). A series of larger autumn storms in late October delivered 3.5 in. of rain that produced the *major* "flush" of the season, mobilizing material accumulated in the soil profile during the dry season; this event was sampled on November 1, 2010 (table 4).

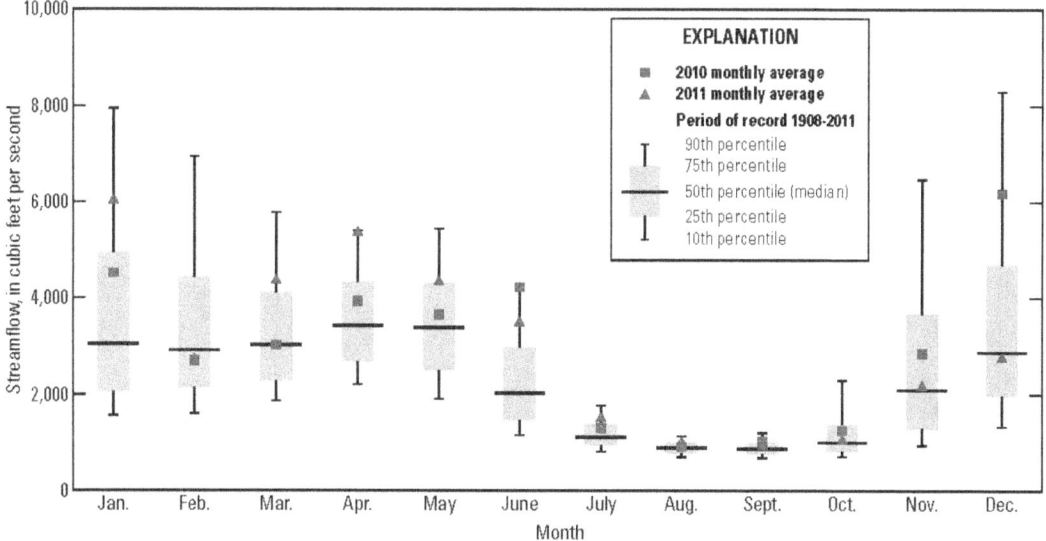

Figure 6. Relation between the average monthly streamflow in the Clackamas River at Estacada, Oregon, in 2010 and 2011 and the 103-year period of record (1908–2011). (Streamflow data from the USGS gaging station 14210000 located at RM 23.1).

Table 6. Summary statistics for select streamflow, water quality, optical properties, and disinfection by-products for the upper and lower main-stem Clackamas River, tributaries, and source water, Clackamas River basin, Oregon.

[Source water includes samples from the Clackamas River Water and City of Lake Oswego drinking-water treatment plants. **Abbreviations:** n, number of samples; Min, minimum; Max, maximum; nm, nanometers; UV, ultraviolet; DBP, disinfection by-product; FP, formation potential; F, filtered; U, unfiltered]

Abbreviation	Constituent definition, units	Upper main stem				Tributaries				Lower main stem and source water				Source water			
		n	Min	Median	Max	n	Min	Median	Max	n	Min	Median	Max	n	Min	Median	Max
Flow	Streamflow, in cubic feet per second	20	343	1,720	11,421	15	0	25	422	65	800	2,247	18,142	38	800	2,269	18,142
Temp	Temperature, in degrees Celsius	20	3.8	9.3	13.4	15	8.0	14.8	19.3	62	4.7	11.2	18.4	38	5.4	9.6	18.4
SC	Specific conductance, in microsemens per centimeter	20	30.0	50.0	82.0	15	32.0	73.0	177.0	62	30.0	50.0	69.0	38	30.0	47.0	69.0
DO	Dissolved oxygen, in milligrams per liter	20	10.5	11.7	13.3	15	9.6	10.3	12.1	62	8.9	11.1	13.0	38	8.9	11.4	12.7
DO%	Percent saturation dissolved oxygen	20	101	104	106	15	96	101	111	62	94	100	112	38	95	100	107
pH	pH, in standard units	20	7.2	7.7	8.0	15	6.8	7.6	8.1	62	7.3	7.6	8.6	38	7.3	7.6	7.8
Turbidity	Turbidity, in formazin nephelometric units	19	0.1	1.7	43	15	0.2	4.1	76.5	62	0.1	1.5	68	38	0.1	1.3	68
Chla (water column)	Chlorophyll-a, in micrograms per liter	20	0.2	0.4	1.0	15	0.4	2.2	7.7	64	0.1	1.1	6.1	37	0.2	1.4	6.1
Chla (benthic)	Chlorophyll-a, in milligrams per square meter	2	52	--	58	10	38	157	304	8	93	201	759	--	--	--	--
SRP	Dissolved phosphate, in milligrams per liter	19	0.010	0.018	0.021	15	0.004	0.031	0.106	64	0.004	0.013	0.030	37	0.004	0.013	0.030
TP	Total phosphorus, in milligrams per liter	18	0.006	0.010	0.033	15	0.009	0.037	0.252	60	0.009	0.010	0.062	36	0.009	0.010	0.062
Si	Dissolved silicate, in milligrams per liter	19	0.7	6.4	9.9	15	2.8	8.5	13.1	64	0.8	6.7	8.9	37	0.9	6.2	8.9
NO3	Dissolved nitrate plus nitrite, in milligrams per liter	19	0.003	0.009	0.025	15	0.059	0.587	1.728	64	0.003	0.026	0.351	37	0.008	0.054	0.351
TPN	Total particulate nitrogen, in milligrams per liter	20	0.009	0.017	0.036	15	0.005	0.029	0.596	65	0.003	0.029	0.177	38	0.006	0.037	0.177
DOC	Dissolved organic carbon, in milligrams per liter	20	0.6	1.1	1.9	15	1.2	2.0	7.8	65	0.8	1.2	2.3	38	0.9	1.3	2.2
TPC	Total particulate carbon, in milligrams per liter	20	0.1	0.2	0.9	15	0.1	0.5	6.3	65	0.1	0.3	2.0	38	0.1	0.3	2.0
TOC	Total organic carbon, in milligrams per liter	20	0.8	1.3	2.8	15	1.3	2.5	13.5	65	0.9	1.5	3.6	38	1.0	1.5	3.6
% C as POC	Particulate carbon, in percent	20	7	17	33	15	5	20	47	65	5	20	55	38	10	23	55

Table 6. Summary statistics for select streamflow, water quality, optical properties, and disinfection by-products for the upper and lower main-stem Clackamas River, tributaries, and source water, Clackamas River basin, Oregon.—Continued

[Source water includes samples from the Clackamas River Water and City of Lake Oswego drinking-water treatment plants. **Abbreviations:** *n*, number of samples; Min, minimum; Max, maximum; nm, nanometers; UV, ultraviolet; DBP, disinfection by-product; FP, formation potential; F, filtered; U, unfiltered]

Abbreviation	Constituent definition, units	Upper main stem				Tributaries				Lower main stem and source water				Source water			
		n	Min	Median	Max	*n*	Min	Median	Max	*n*	Min	Median	Max	*n*	Min	Median	Max
FDOM	Fluorescence FDOM, in water Raman units	20	0.02	0.04	0.09	15	0.05	0.14	0.49	65	0.02	0.04	0.12	38	0.03	0.05	0.11
PEAK A	Fluorescence Peak A, in water Raman units	20	0.03	0.09	0.21	15	0.10	0.33	1.23	65	0.05	0.10	0.25	38	0.07	0.10	0.25
PEAK C	Fluorescence Peak C, in water Raman units	20	0.02	0.05	0.11	15	0.06	0.19	0.62	65	0.03	0.05	0.14	38	0.04	0.06	0.14
PEAK M	Fluorescence Peak M, in water Raman units	20	0.01	0.04	0.08	15	0.05	0.15	0.60	65	0.03	0.04	0.12	38	0.03	0.05	0.12
PEAK D	Fluorescence Peak D, in water Raman units	20	0.01	0.03	0.06	15	0.03	0.08	0.30	65	0.01	0.03	0.08	38	0.02	0.03	0.07
PEAK B	Fluorescence Peak B, in water Raman units	20	0.02	0.03	0.12	15	0.03	0.09	0.52	65	0.02	0.04	0.10	38	0.02	0.04	0.10
PEAK T	Fluorescence Peak T, in water Raman units	20	0.02	0.04	0.10	15	0.04	0.10	0.57	65	0.02	0.04	0.14	38	0.03	0.04	0.14
PEAK N	Fluorescence Peak N, in water Raman units	20	0.01	0.03	0.07	15	0.04	0.14	0.56	65	0.02	0.04	0.13	38	0.03	0.04	0.13
FI	Fluorescence Index, unitless	20	1.33	1.38	1.46	15	1.39	1.44	1.50	65	1.31	1.38	1.45	38	1.34	1.38	1.45
HIX	Humic Index, unitless	20	2.13	3.49	6.08	15	2.10	4.71	7.52	65	2.36	3.68	5.32	38	2.36	3.69	4.75
C1 Loading	Component 1, in fluorescence maximum values	20	0.01	0.04	0.09	15	0.04	0.14	0.94	65	0.02	0.04	0.17	38	0.03	0.05	0.17
C2 Loading	Component 2, in fluorescence maximum values	20	0.01	0.03	0.08	15	0.04	0.09	0.35	65	0.02	0.03	0.10	38	0.02	0.04	0.10
C3 Loading	Component 3, in fluorescence maximum values	20	0.02	0.05	0.12	15	0.06	0.18	0.59	65	0.03	0.05	0.14	38	0.04	0.06	0.14
C4 Loading	Component 4, in fluorescence maximum values	20	0.02	0.04	0.10	15	0.05	0.16	0.52	65	0.03	0.05	0.12	38	0.03	0.05	0.12
C5 Loading	Component 5, in fluorescence maximum values	20	0.01	0.03	0.09	15	0.02	0.08	0.56	65	0.02	0.03	0.10	38	0.02	0.03	0.10
Total C Loading	Total component, in fluorescence maximum values	20	0.08	0.18	0.45	15	0.24	0.72	2.81	65	0.13	0.21	0.57	38	0.15	0.22	0.57

Table 6. Summary statistics for select streamflow, water quality, optical properties, and disinfection by-products for the upper and lower main-stem Clackamas River, tributaries, and source water, Clackamas River basin, Oregon.—Continued

[Source water includes samples from the Clackamas River Water and City of Lake Oswego drinking-water treatment plants. **Abbreviations:** n, number of samples; Min, minimum; Max, maximum; nm, nanometers; UV, ultraviolet; DBP, disinfection by-product; FP, formation potential; F, filtered; U, unfiltered]

Abbreviation	Constituent definition, units	Upper main stem				Tributaries				Lower main stem and source water				Source water			
		n	Min	Median	Max	n	Min	Median	Max	n	Min	Median	Max	n	Min	Median	Max
% C1	Percentage of component 1	20	13	19	22	15	16	22	41	65	17	21	33	38	18	21	33
% C2	Percentage of component 2	20	13	16	20	15	11	14	18	65	11	16	21	38	11	17	19
% C3	Percentage of component 3	20	20	26	29	15	19	25	29	65	16	25	30	38	19	26	28
% C4	Percentage of component 4	20	18	22	25	15	17	22	26	65	18	22	24	38	18	23	24
% C5	Percentage of component 5	20	8.3	17	31	15	5.4	12	30	65	5.7	16	27	38	8.2	15	23
$S_{275-295}$	Spectral Slope for 275-295 nm, in per nm	20	0.01	0.01	0.02	15	0.01	0.01	0.01	65	0.01	0.01	0.02	38	0.01	0.01	0.02
$S_{290-350}$	Spectral Slope for 290-350 nm, in per nm	20	0.01	0.01	0.02	15	0.01	0.01	0.02	65	0.01	0.01	0.03	38	0.01	0.01	0.03
$S_{350-400}$	Spectral Slope for 350-400 nm, in per nm	19	0.01	0.01	0.02	15	0.01	0.01	0.01	63	0.01	0.01	0.02	37	0.01	0.01	0.02
SR	Spectral Slope Ratio, unitless	19	0.67	0.96	1.50	15	0.78	0.92	1.23	63	0.71	0.98	2.05	37	0.71	0.98	2.05
A_{254}	Absorbance at 254 nm, in per centimeter	20	0.01	0.03	0.07	15	0.03	0.05	0.39	65	0.01	0.03	0.10	38	0.01	0.03	0.10
SUVA	Specific UV absorbance at 254 liters per milligram per meter	20	1.43	2.60	4.06	15	2.00	2.88	4.03	65	1.64	2.64	4.41	38	2.00	2.73	4.41
THMFP-F	DBPFP for total trihalomethanes, in milligrams per liter (filtered)	5	0.055	0.096	0.130	15	0.113	0.198	0.595	20	0.091	0.134	0.158	8	0.119	0.145	0.156
HAAFP-F	DBPFP for total haloacetic acids, in milligrams per liter (filtered)	5	0.010	0.015	0.059	15	0.030	0.071	0.637	20	0.016	0.026	0.076	8	0.020	0.031	0.064
THMFP-U	DBPFP for total trihalomethanes, in milligrams per liter (unfiltered)	5	0.063	0.083	0.123	15	0.116	0.194	0.956	20	0.096	0.148	0.237	8	0.140	0.166	0.223
HAAFP-U	DBPFP for total haloacetic acids, in milligrams per liter (unfiltered)	5	0.013	0.038	0.096	15	0.036	0.087	0.929	20	0.021	0.040	0.154	8	0.025	0.044	0.154

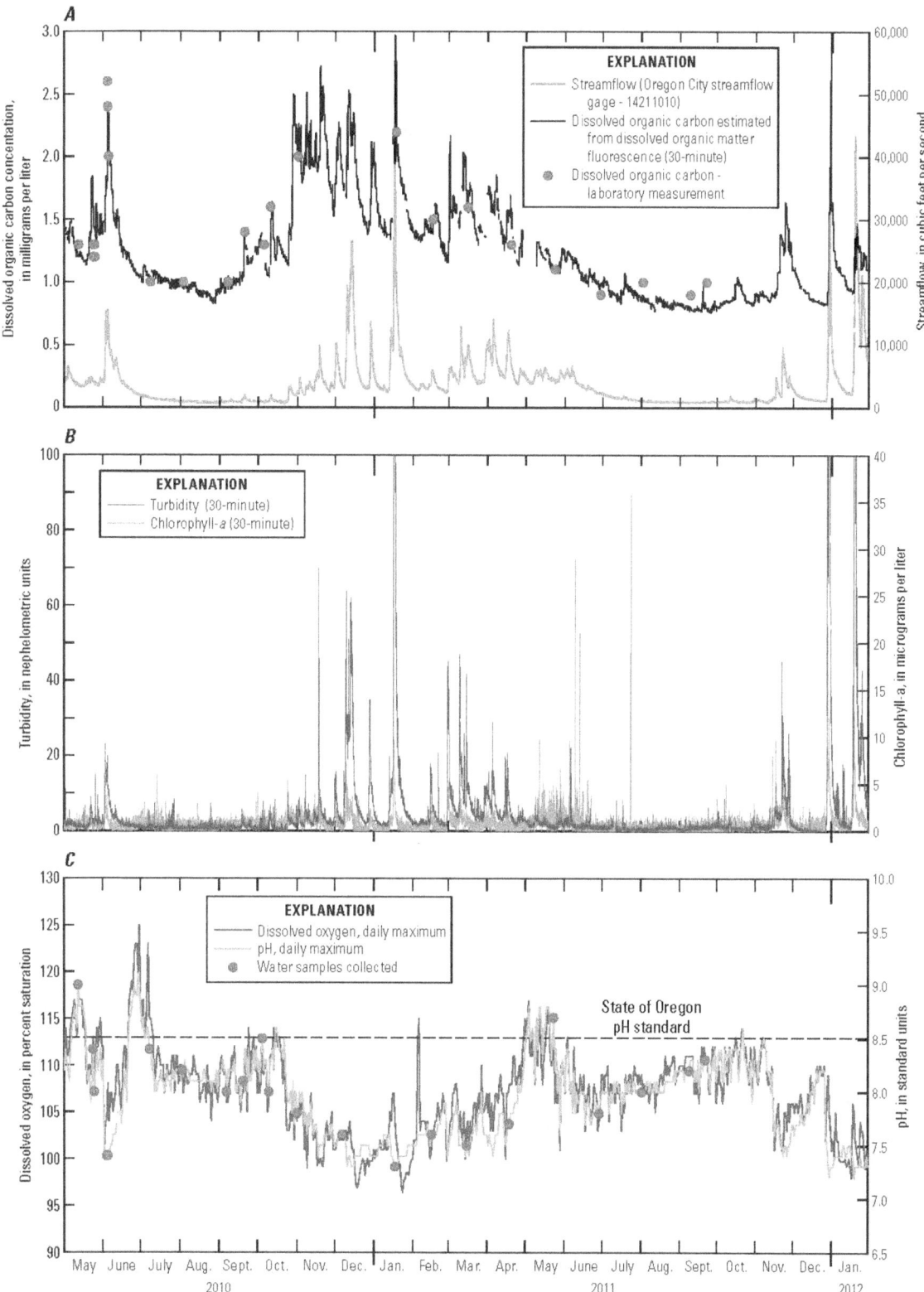

Figure 7.   Time series of (*A*) streamflow and dissolved organic carbon, (*B*) turbidity and chlorophyll-*a*, and (*C*) dissolved oxygen and pH in the lower Clackamas River, Oregon, 2010–11.

The 2011 water year and growing season were similarly wet and cool; higher-than-average streamflow resulted in relatively low water-column concentrations of chlorophyll-*a* (fig. 8). *Anabaena* bloomed again in September 2011, leading to another advisory for North Fork Reservoir and Timothy Lake (fig. 1) on September 8, 2011. The bloom in North Fork Reservoir was not as severe as in years past (Carpenter, 2003), although the bloom did produce enough biomass to form a surface scum on the reservoir and may have contributed to reports of tastes and odors in treated drinking water. In addition to the depth-profile sampling at the log boom in North Fork Reservoir, one surface sample containing a high abundance of *Anabaena* was collected at Promontory Park during the bloom (see photograph 2b). This sample represents DOM enriched in algal-derived carbon and served as an "end member" in terms of carbon characterization.

The third and fourth basin-wide surveys were conducted near base-flow conditions in early and mid-September, respectively, a low-carbon period for the river. The final sampling was during the annual drawdown of Timothy Lake (table 4). In addition to the typical suite of main-stem sites, the fourth survey included the main-stem site at the Two Rivers Campground (fig. 1), a site not influenced by the Timothy Lake release. Although in years past the Timothy Lake drawdown contributed higher concentrations of DOC, TOC, chlorophyll-*a*, and *Anabaena* cells to the upper mainstem (Carpenter, 2003), in 2011, the observed effect was limited to a three-fold increase in TOC at Carter Bridge that continued downstream to the DWTP intakes. Although discrete sampling ended with this last synoptic, the in-situ FDOM sensors continued to operate through January 2012.

Algal growth in the Clackamas River and its impact on diel changes in pH and DO are affected by factors such as streamflow, water temperature, nutrients, and availability of solar radiation for photosynthesis. The higher flows during this study, especially during springtime (fig. 6), produced lower-than-average water temperatures and probably also delayed colonization by benthic algae in faster velocity zones. These conditions, exacerbated by clouds and scant sunshine, delayed and (or) truncated the algal growing season both years. While much higher-than-average rainfall and prolonged cloudy weather in 2010 might have resulted in less solar radiation available for growing periphyton, field surveys in June 2010 revealed areas of high algal biomass, especially in the lower mainstem where periphyton chlorophyll-*a* levels exceeded the commonly applied nuisance threshold of 100–150 mg/m^2 (Welch and others, 1988) at all four main-stem sites—Estacada, Barton, Carver, and Highway 99E (table 7). By early September, periphyton biomass was even higher at the tributary sites but lower at all lower-basin main-stem sites, possibly from slow, metered losses due to the scouring high flows, grazing by benthic invertebrates, or some other factor.

Despite the apparent reduction in benthic algal biomass in the lower mainstem, conditions either precluded or obscured the occurrence of any substantial sloughing of viable chlorophyll-*a* into the water column, or the material was not particularly prone to fluoresce. In fact, concentrations of chlorophyll-*a* in the water column remained low and, on average, were much lower compared with previous years (fig. 8). It is possible that higher flow, as well as higher water velocity, may have produced a more steady (and less punctuated) losses of periphyton algal particles into the water column.

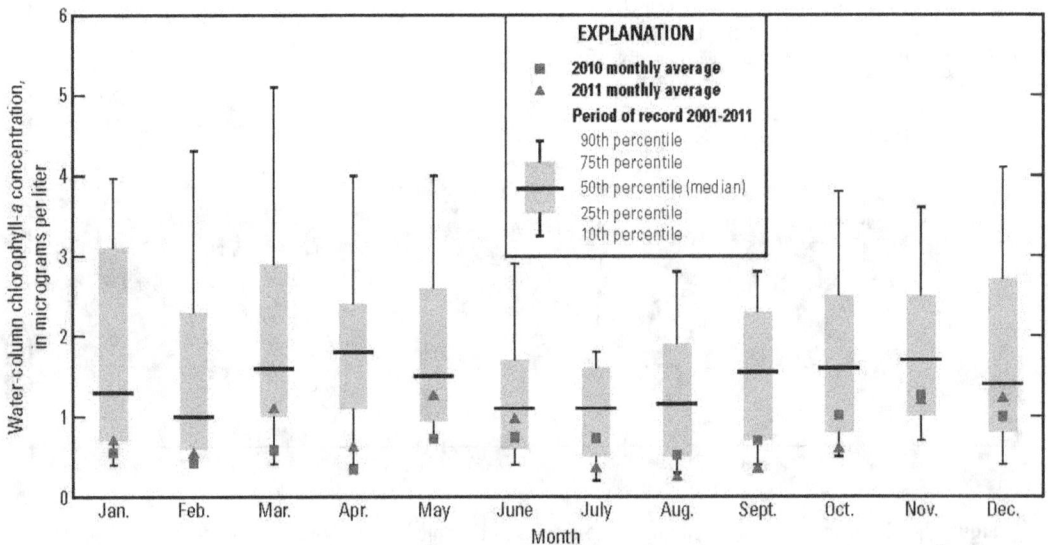

Figure 8. Relation between the average monthly water-column chlorophyll-*a* concentrations in the Clackamas River at Oregon City, Oregon, in 2010 and 2011, and the 9-year period of record (2001–11).

Table 7. Benthic algal conditions in the Clackamas River and select tributaries, Oregon.

[Sampling site locations are shown in figure 1. Algal biomass given as concentration of chlorophyll-a in milligrams per square meter. **Abbreviation:** sp., species]

Sampling sites	Benthic algae description	Benthic algal biomass
July–August 2010		
Clackamas River at Carter Bridge	Green algae (*Prasiola* sp., *Ulothrix* sp., *Zygnema* sp.), stalked diatoms (Cymbella mexicana), benthic diatoms (Epithemia sp.), and blue-green algae (*Nostoc* sp.)	52
Clackamas River at Estacada	Stalked diatoms (*Cymbella mexicana*), benthic diatoms (*Melosira* sp. and *Epithemia* sp.)	421
Clackamas River at Barton Bridge	Filamentous green algae (*Cladophora* sp.), stalked diatoms (*Cymbella mexicana*), benthic diatoms (*Melosira, Gomphonema, Synedra,* and *Nitzschia* sp., *Epithemia* sp.), and lesser amounts of filamentous blue-green algae (*Oscillatoria* sp.)	251
Clackamas River at Carver	Stalked diatoms (*Cymbella mexicana*), benthic diatoms (*Melosira* sp., *Gomphonema* sp.), with lesser amounts of filamentous green algae (*Cladophora* sp.)	759
Clackamas River upstream of Highway 99E	Filamentous green algae (*Cladophora* sp.), stalked diatoms (*Cymbella mexicana*), benthic diatoms (*Melosira, Gomphonema, Synedra,* and *Nitzschia* sp.), and red algae (*Lemanea* sp.)	342
Eagle Creek	Stalked diatoms (*Gomphoneis* sp.), and blue-green algae (*Oscillatoria* sp. and Rivulariceaen heterocystous filaments)	47
Deep Creek	Benthic diatoms (*Melosira* sp., *Synedra* sp., *Navicula* sp., and *Cocconeis* sp.)	120
Clear Creek	Benthic diatoms (*Melosira* sp., *Gomphonema* sp., *Synedra* sp., *Navicula* sp., and *Cocconeis* sp.)	38
Rock Creek	Benthic diatoms (*Melosira* sp., *Rhoicosphenia* sp., *Synedra* sp., *Navicula* sp., *Nitzschia* sp., and *Cocconeis* sp.)	181
Sieben Creek	Benthic diatoms (*Melosira* sp., *Synedra* sp., *Achnanthidium* sp., *Navicula* sp., *Nitzschia* sp., and *Cocconeis* sp.) and green algae (*Closterium* sp.)	155
September 2010		
Clackamas River at Carter Bridge	Filamentous green algae (*Ulothrix* sp.), blue-green algae (*Nostoc* sp.), stalked diatoms (*Cymbella mexicana*), and benthic diatoms (*Epithemia* sp. and *Rhopalodia* sp.)	58
Clackamas River at Estacada	Stalked diatoms (*Cymbella mexicana*), benthic diatoms (*Epithemia* sp., *Synedra* sp., *Gompohonema* sp.)	152
Clackamas River at Barton Bridge	Filamentous green algae (*Cladophora* sp.), stalked diatoms (*Cymbella mexicana*), benthic diatoms (*Melosira, Gomphonema, Synedra,* and *Rhopalodia* sp.)	144
Clackamas River at Carver	Filamentous green algae (*Cladophora* sp.), benthic diatoms (*Melosira, Gomphonema, Synedra,* and *Rhopalodia* sp.), and filamentous blue-green algae (*Oscillatoria* sp.)	125
Clackamas River upstream of Highway 99E	Filamentous green algae (*Cladophora* sp.), benthic diatoms (*Melosira, Gomphonema, Synedra,* and *Cocconeis* sp.), and red algae (*Lemanea* sp.)	93
Eagle Creek	Blue-green algae (*Oscillatoria* sp.) and benthic diatoms (*Gomphonema* sp., *Cymbella minuta, Epithemia* sp.)	84
Deep Creek	Benthic diatoms (*Melosira* sp., *Synedra* sp., *Navicula* sp.) and green algae (*Closterium* sp.)	193
Clear Creek	Benthic diatoms (*Melosira* sp., *Synedra* sp., *Navicula* sp., and *Cymbella* sp.)	159
Rock Creek	Benthic diatoms (*Melosira* sp., *Synedra* sp., *Navicula* sp., *Nitzschia* sp., and *Rhoicosphenia* sp.)	304
Sieben Creek	Benthic diatoms (*Melosira* sp. and *Bacillaria paradoxa*) and filamentous blue-green algae (*Oscillatoria* sp.)	301

Dissolved and Particulate Carbon Concentrations

DOC concentrations were generally low in the Clackamas River, typically about 1.0–1.5 mg/L; during storms, concentrations occasionally increased up to about 2.5 mg/L (fig. 9). DOC concentrations were much higher in the tributaries during the one storm sampling in October 2010; Rock Creek had the highest concentration (7.8 mg/L). Within the mainstem, carbon concentrations (DOC and TOC) increased downstream to the drinking-water intakes (fig. 10). Longitudinal increases between the Carter Bridge and Estacada sites, although not large, may be attributed to an effect of the hydroelectric project reservoirs, including North Fork Reservoir, but also could be caused by input from Wade Creek (fig. 1), which drains the city of Estacada and receives treated effluent from the wastewater treatment facility that discharges to the Clackamas River upstream from River Mill dam (fig. 3).

There was a strong correlation between concentrations of DOC and TOC ($r = 0.98$, $p < 0.001$). In general, the TOC pool was dominated by the dissolved fraction (greater than 70 percent); there were, however, times when the particulate fraction (TPC) was significant, making up 30 to 40 percent of the TOC (fig. 10). In the tributaries, the highest percentage of carbon as TPC was during the October 10, 2010, initial storm event when turbidity also was relatively high. The TPC fraction of the TOC pool was also high (approximately 40 percent) in several of the samples from North Fork Reservoir in 2010. In mid-March 2011, an unusually high TPC value in the source-water sample from the CRW DWTP intake (2.0 mg/L) resulted in a relatively high TOC value (3.6 mg/L) that was largely made up of TPC (55 percent). Given that just 2.2 mi downstream at the LO DWTP, TPC was 0.5 mg/L and TOC was 2.0 mg/L that day, this relatively high TPC value at CRW DWTP was likely the result of resuspension of particles accumulated within the intake vault rather than being reflective of river conditions.

Seasonal variations in streamflow had a pronounced effect on turbidity and carbon concentrations in the mainstem and tributaries and on DBPs in finished water. The DOC concentrations in source water were highest (2.0 to 2.6 mg/L) in June and November 2010 and January 2011—all samples affected by storm runoff (fig. 10). Carbon concentrations were commonly closely tied to streamflow (fig. 7) and resulted in a significant positive correlation between streamflow and DOC ($r = 0.82$, $p < 0.001$). The effect was, however, complex because higher streamflow also occasionally diluted carbon concentrations. The net effect of this situation is that over the course of a storm, or longer periods of time with successive storms, there is a "hysteresis effect." Relative to streamflow, carbon concentrations are initially high as the storms mobilize organic matter from the watershed, but as the storm continues, concentrations are lower, relative to streamflow, because less carbon is flushed later, leading to lower concentrations from the effect of greater dilution (Pellerin and others, 2012).

Disinfection By-Product Concentrations in Finished Drinking Water

DBP concentrations in finished water from the CRW and LO DWTPs were similar, about 0.024 ± 0.006 mg/L for THM4 and 0.022 ± 0.008 mg/L for HAA5 (fig. 11). Maximum DBP concentrations in finished water also were similar for the two classes of DBPs, about 0.04 mg/L for both THM4 and HAA5. Concentrations of THM4 and HAA5 for the two samples collected from within the distribution systems were, however, higher (fig. 11).

Following the winter high-flow period, streamflow in the Clackamas River declined through summer as the snow packs diminished and groundwater made up a progressively greater contribution to the flow in the river. This seasonal transition for the Clackamas River and other rivers draining the Cascade Range was first described by Piper (1942). The lower turbidity and carbon levels that resulted (fig. 7) produced lower concentrations of DBPs in finished water (fig. 11). Lower concentrations of DBPs during a time when (regional) groundwater dominates flow is consistent with findings from a USEPA study (U.S. Environmental Protection Agency, 2005) that reported generally lower concentrations of THMs in finished water when source water was from groundwater rather than from surface-water sources.

The seasonal patterns in concentrations of THM4 and HAA5 and overall composition of DBPs in finished water also were similar at the two DWTPs. Chloroform was the dominant THM in finished water, making up 86–97 percent of the THM4 for both DWTPs; bromodichloromethane made up the remaining 3–14 percent. For HAAs, DCAA and TCAA were the primary DBPs, making up 40–60 percent each. MBAA was detected in finished water from both DWTPs at concentrations equal to the detection limit of 0.001 mg/L; no other DBPs were detected.

An examination of individual concentrations and Benchmark Quotient (BQ) values can reveal times when concentrations are most elevated, and how close they may be to existing standards. BQ values indicated no drinking-water standards were exceeded (fig. 12). The maximum BQ values were somewhat higher for HAA5, which has a lower MCL (0.060 mg/L) than THM4 (0.080 mg/L). Drinking-water standards for DBPs are based on the annual running average of the maximum concentrations from within the distribution system, not the "time zero" finished-water samples collected from within the treatment plant. The quarterly monitoring data for the CRW and LO DWTPs showed similar maximum BQ values—about 0.35 for THM4 and 0.46 for HAA5 (Oregon Health Authority, 2012) as those reported here. The two samples collected from within the distribution system (one from each DWTP) had the highest BQs of 0.76 and 0.93, which indicate that HAA5 concentrations approached but did not exceed USEPA standards.

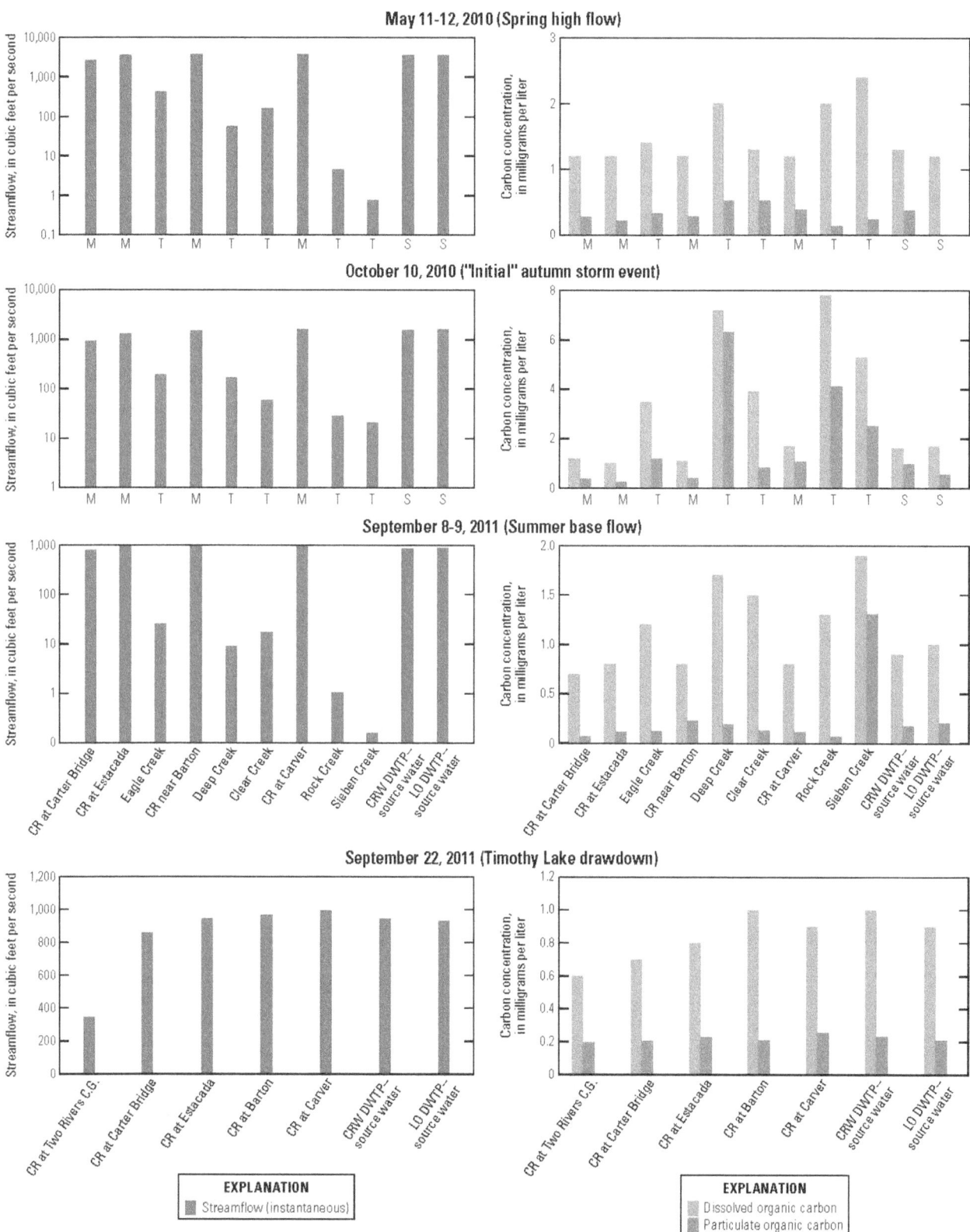

Figure 9. Streamflow and concentrations of dissolved and particulate organic carbon in the tributaries and main-stem Clackamas River, Oregon, 2010–11. (Sites are listed in downstream order. Note variable y-axis scales, and log scale for y-axis in streamflow plots. X-axis labels for Summer base flow also apply to the Spring high flow and Initial autumn storm event. Abbreviations: M, main-stem site; T, tributary; S, source-water intake; mg/L, milligram per liter; CR, Clackamas River; CRW, Clackamas River Water DWTP; LO, City of Lake Oswego DWTP; DWTP, drinking-water treatment plant; C.G., campground.)

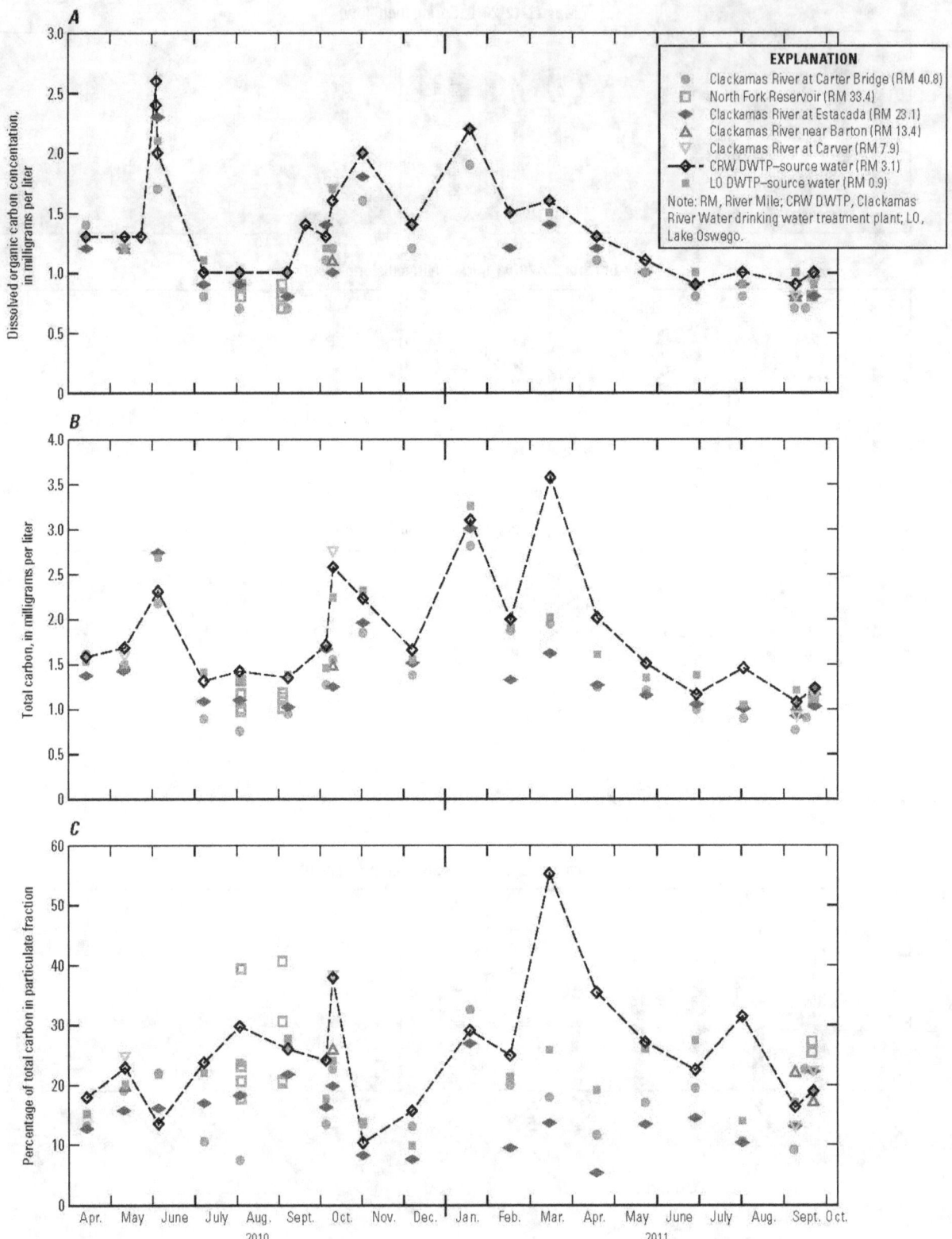

Figure 10. Seasonal patterns in (*A*) dissolved organic carbon, (*B*) total carbon, and (*C*) percentage of total carbon in particulate fraction in the main-stem Clackamas River, Oregon, 2010–11.

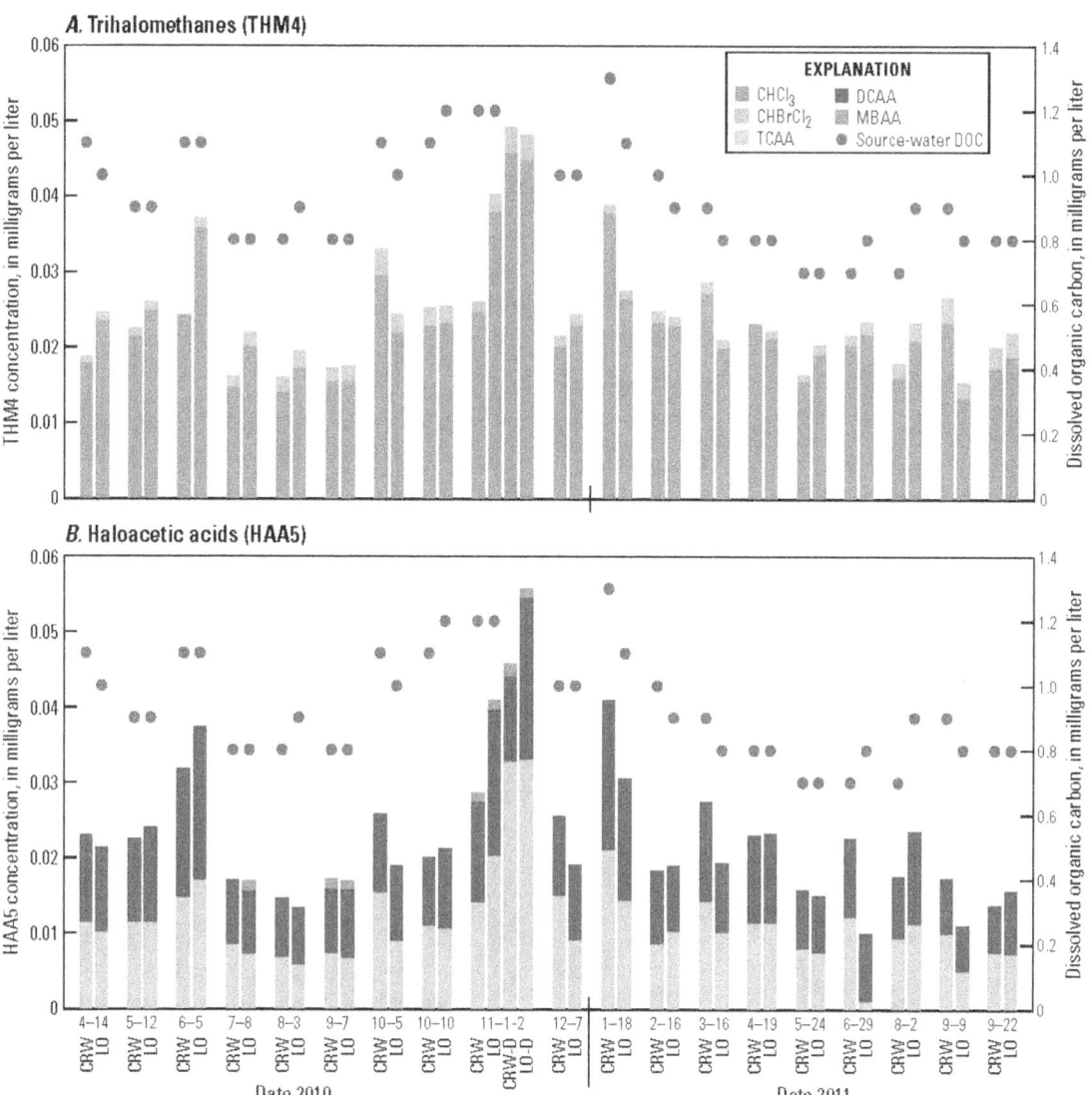

Note: CHCl₃, chloroform; CHBrCl₂, bromodichloromethane; TCAA, trichloroacetic acid; DCAA, dichloroacetic acid; MBAA, monobromoacetic acid; CRW, Clackamas River Water; LO, City of Lake Oswego; D, distribution sample; DOC, dissolved organic carbon; THM4, trihalomethanes; HAA5, haloacetic acids.

Figure 11.   Seasonal patterns in (*A*) total trihalomethanes (THM4) and (*B*) total haloacetic acids (HAA5) in finished water from the Clackamas River Water (CRW) and City of Lake Oswego (LO) drinking-water treatment plants, Clackamas River basin, Oregon, 2010–11.

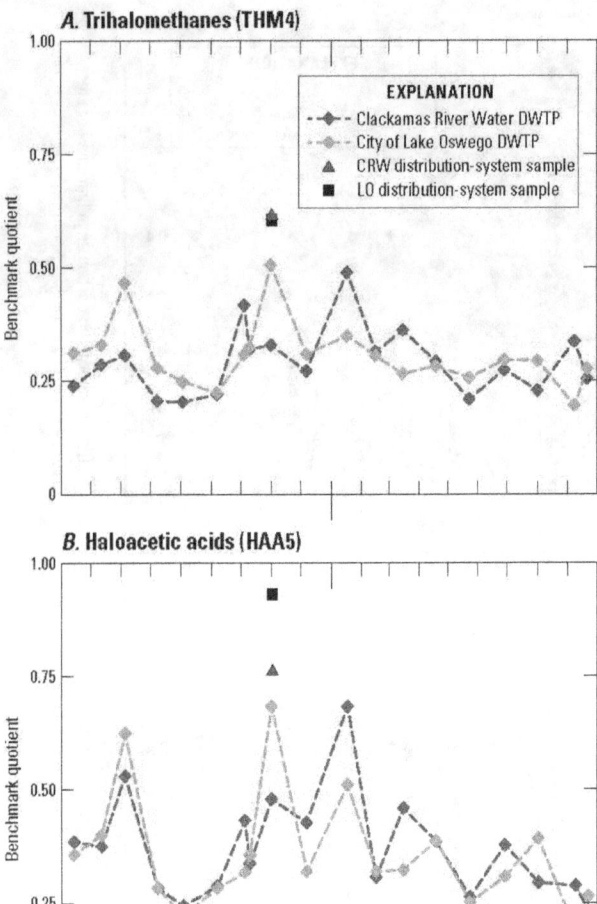

A. Trihalomethanes (THM4)

EXPLANATION
◆ Clackamas River Water DWTP
◆ City of Lake Oswego DWTP
▲ CRW distribution-system sample
■ LO distribution-system sample

B. Haloacetic acids (HAA5)

Figure 12. Seasonal patterns in disinfection by-product benchmark quotients (concentrations relative to drinking-water standards) for (*A*) four trihalomethanes and (*B*), five haloacetic acids in finished water from the Clackamas River Water and City of Lake Oswego drinking-water treatment plants, Clackamas River basin, Oregon, 2010–11.

Storm Effects on Organic Carbon Concentrations and Disinfection By-Products

Several large storm events during the study caused surface runoff (see photographs 5a-b), including one high-flow event in June 2010 that resulted in a peak DOC concentration of about 2.5 mg/L (fig. 7). In October, with the onset of autumn precipitation, DOC concentration increased about 50 percent in the lower Clackamas River during the initial storm (fig. 13). This storm, the first significant precipitation event in months, caused moderate runoff that increased turbidity in the mainstem from 0.5 to 4.5 FNUs, more than doubling the TPC and TPN at the CRW DWTP intake (table 8). Although a shift in DOM composition was observed at the CRW intake during this storm, it did not produce any notable change in finished-water DBPs (fig. 11).

Photograph 5a-b. Clear Creek upstream from Carver Park running turbid after a storm showing its effect on the main-stem Clackamas River below. (Photographs by Kurt Carpenter, U.S. Geological Survey, November 14, 2011 [top] and May 25, 2010 [bottom].)

Photograph 5b.

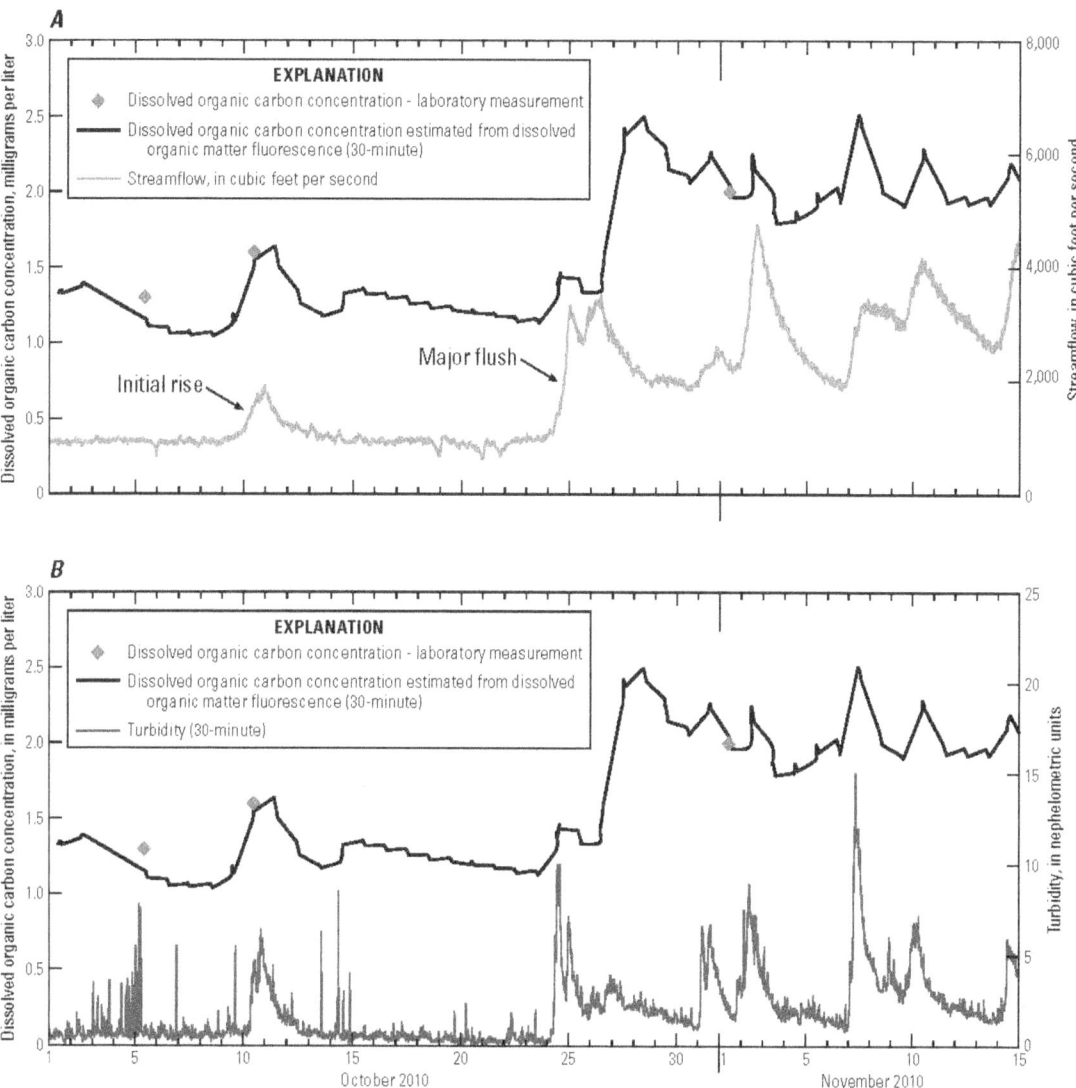

Figure 13. Streamflow, turbidity, and dissolved organic carbon in the lower Clackamas River at the Clackamas River Water drinking-water treatment plant intake, October–November 2010, highlighting two different responses to storms.

The second, larger storm at the end of October 2010 (sampled on November 1st) was more characteristic of a true soil "flushing" that mobilized carbon from the watershed into the hydrologic system. Although turbidity was not as high during this second storm, the DOC increased to 2 mg/L at the DWTPs, or about 2.5 times higher than during the summer base-flow period (table 8 and fig. 7). More importantly, the DOC was reactive, producing the highest concentrations of DBPs in finished water during this study: 0.04–0.05 mg/L

THM4 and 0.03–0.04 mg/L HAA5. The two samples collected from within the distribution system had higher concentrations (fig. 11). In addition to increased DBPs, moderate taste and odor issues began at this time, along with increased demands for chlorine and coagulant doses (Kari Duncan, City of Lake Oswego, written commun., 2011). Even though the first storm mobilized quantities of carbon, concentrations of DBPs were higher in finished water later in the autumn after the ground was saturated.

Table 8. Trends in source-water and finished-water quality during two storms in the Clackamas River basin, Oregon, October–November 2010 at the Clackamas River Water drinking-water treatment plant.

[For ease in comparision among sampling dates, red and black arrows indicate trends in constituent values. **Abbreviations:** mg/L, milligram per liter; µg/L, microgram per liter; nm, nanometers]

Parameter	Pre-storm (October 5, 2010)	Trend	"Initial" autumn storm event (October 10, 2010)	Trend	"Major flush" event (November 1, 2010)
Clackamas River Water—source water					
Streamflow, in cubic feet per second	951	↑	1,548	↑	2,247
Turbidity, in nephelometric units	0.5	↑	4.5	↓	2.1
Dissolved organic carbon, in mg/L	1.3	↑	1.6	↑	2.0
Total particulate carbon, in mg/L	0.41	↑	0.98	↓	0.23
Total particulate nitrogen, in mg/L	0.06	↑	0.11	↓	0.04
Chlorophyll-*a* (water column), in µg/L	1.7	↑	3.8	↓	2.3
Specific ultraviolet absorbance at 254 liters per milligram per meter	2.08	↑	2.75	↑	3.25
Fluorescence index	1.41	↑	1.44	↓	1.40
Humic index	3.52	↑	3.61	↑	4.29
Percentage of component C1	19	↑	29	↓	18
Percentage of component C2	13	↓	12	↑	18
Percentage of component C3	24	↓	21	↑	28
Percentage of component C4	22	↓	20	↑	24
Percentage of component C5	22	↓	17	↓	12
Clackamas River Water—finished water					
Total trihalomethanes (THM4)	0.033	↓	0.025	↑	0.026
Total haloacetic acids (HAA5)	0.026	↓	0.020	↑	0.029

Disinfection By-Product Formation Potentials

During each basin-wide survey (table 4), laboratory measurements of DBPFP were made to compare the propensity for waters from a variety of locations within the watershed to form DBPs upon chlorination. These instantaneous concentrations, and the load and yield calculations based on these concentrations, reflect stream conditions at the time of sampling; because the discrete sampling only captured one point in time, there likely are considerable variations in these estimates. These variations were probably limited to periods of dynamic river conditions, especially during the October 2010 storm; data from the three other basin-wide surveys, conducted during stable streamflow, were less affected by such temporal variability. Additionally, the DBPFP values represent DBPs formed from non-coagulated water under controlled laboratory conditions and do not necessarily equate with DBPs formed during actual water treatment.

By far, the highest DBPFP values were from the tributaries (fig. 14), particularly Deep, Rock, and Sieben Creeks, where DOC concentrations were also highest.

Concentrations were particularly high—up to almost 1.0 mg/L in Deep Creek—during the October 10, 2010, storm event, when longitudinal increases in main-stem DBPFP values were observed. These high concentrations are likely the result of these tributary inputs.

DBPFPs were measured on filtered and unfiltered water, allowing for an evaluation of the relative importance of particles (greater than 0.7 µm) to total DBP formation. Particulate carbon occurs in the form of sediment, detritus, and plant cells including floating and detached benthic algae. Although the majority of the DBPs that formed (60–100 percent of the THMFP and 40–100 percent of the HAAFP) were attributed to the dissolved fraction, particulate carbon also contributed DBPs. Considering just main-stem sites, on average, 10 percent of the unfiltered THMFP and 32 percent of the unfiltered HAAFP were attributed to particles. A comparison of the formation potentials for filtered and unfiltered samples also generally showed higher values in unfiltered samples for the three primary DBPs in finished water—chloroform, DCAA, and TCAA (fig. 15). This finding suggests there would be some reduction in DBPs by introducing a coagulation step prior to chlorination during water treatment.

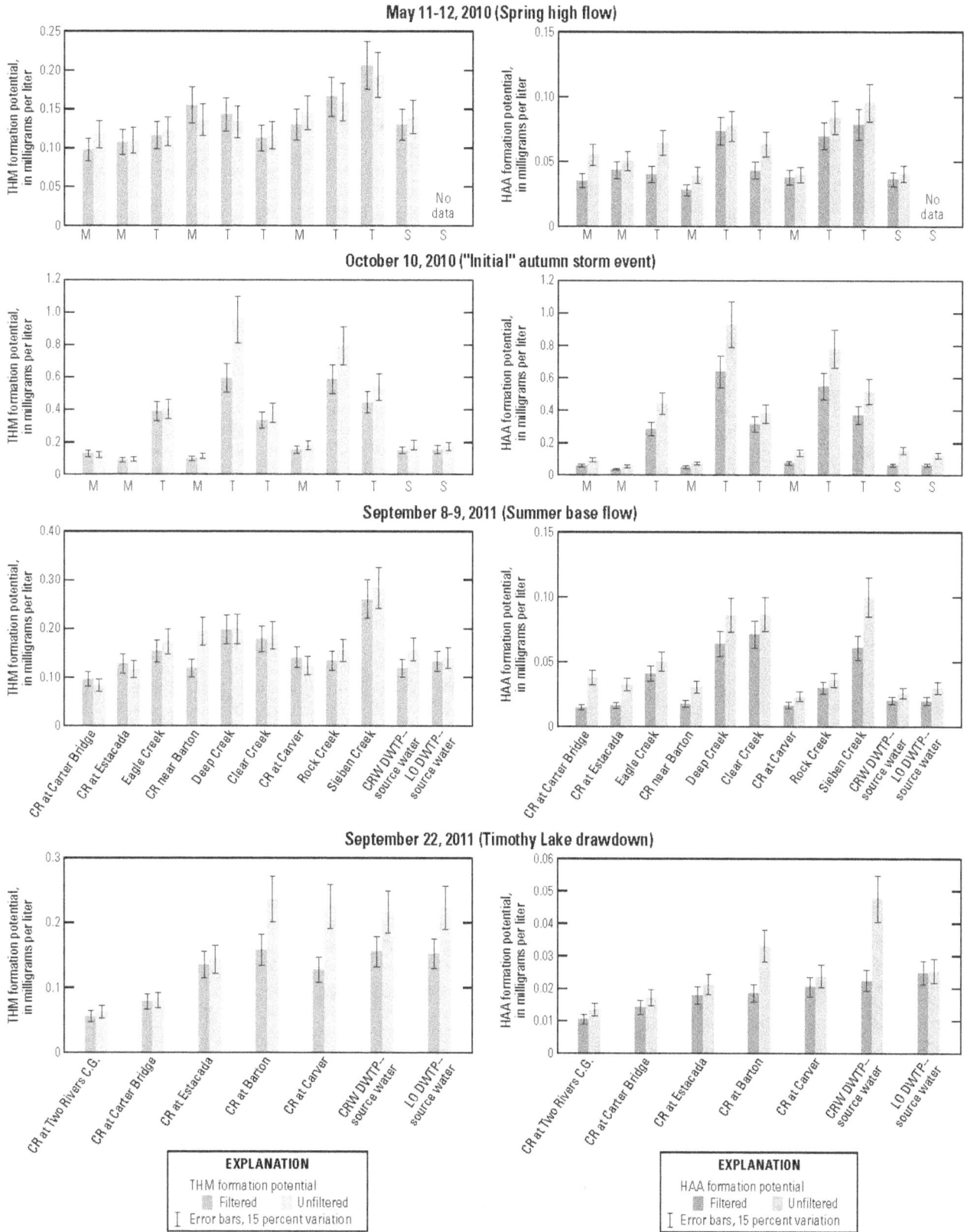

Figure 14. Disinfection by-product formation potentials for total trihalomethanes (THM4) and total haloacetic acids (HAA5) for samples collected in the Clackamas River basin, Oregon, 2010–11. (Sites are listed in downstream order. Note variable Y-axis scales. X-axis labels for Summer base flow also apply to the Spring high flow and Initial autumn storm event. Abbreviations: M, main-stem site; T, tributary; S, source-water; mg/L, milligram per liter; CR, Clackamas River; CRW, Clackamas River Water DWTP; LO, City of Lake Oswego DWTP; DWTP, drinking-water treatment plant; C.G., campground.)

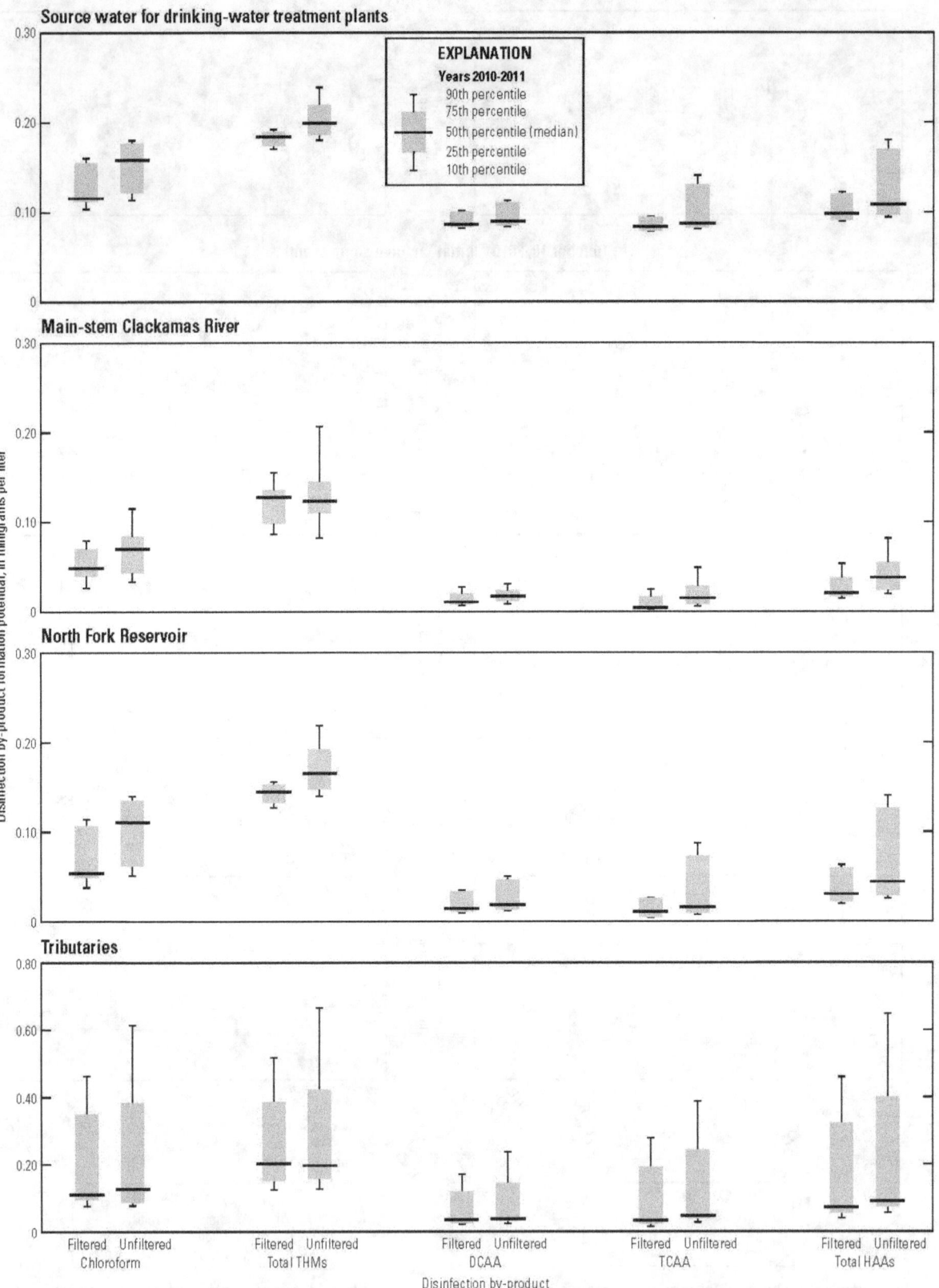

Figure 15. Range of disinfection by-product formation potentials for filtered and unfiltered samples from the Clackamas River basin, Oregon, 2010–11. (Abbreviations: THMs, trihalomethanes; DCAA, dichloroacetic acid; TCAA, trichloroacetic acid; HAA, haloacetic acid.)

It should be noted that although they contribute a relatively small percentage to the total DBPs in treated Clackamas River water, brominated DBPs are of potential concern because of their high molecular weights and potentially more problematic human-health effects compared with chlorinated DBPs. Brominated DBPs form if bromide is present in the source water because it reacts with chlorine and organic matter to form DBPs more quickly than chlorine alone. Although bromide in the Clackamas Basin could, potentially, originate from marine influences (air currents that deliver salts in sea spray), the major source is probably Austin Hot Springs, a hydrothermal suite of springs located along the upper Clackamas River about 3 mi upstream from the Two Rivers Campground sampling site (fig. 1). Chemical analyses conducted 40 years ago showed a bromide concentration of 1.2 mg/L in the spring, which discharged 275 gal/min (U.S. Geological Survey, 2006). The upper Clackamas River at Two Rivers campground had the highest proportion of brominated DBPs during this study, likely reflecting the higher bromide levels at this site compared with downstream locations.

Specific Disinfection By-Product Formation Potentials

STHMFP and SHAAFP values provide information about the degree to which carbon from various locations in the basin reacts with chlorine to form DBPs. Shifts in these values reflect changes in DOM composition and, thus, its source and processing. Surprisingly, there was a significant negative correlation between STHMFP and SHAAFP for both filtered and unfiltered samples (appendixes F5 and F6). This suggests precursor sources for these two classes of DBPs can differ from each other in the Clackamas Basin, as has been reported in other studies (Krasner and others, 2006; Kraus and others, 2008, 2010).

The STHMFP values for filtered samples were highest in September 2011 (fig. 16 and table 9). The highest STHMFP value was associated with the 80-ft depth release point within North Fork Reservoir during a seemingly small blue-green algae bloom. STHMFP values at downstream main-stem sites were also elevated at this time, including source water for the CRW and LO DWTPs. The two basin-wide sampling events at this time represented the summer low-flow period and the seasonal drawdown of Timothy Lake, which contributed about 20 percent of the flow in the lower mainstem. The two STHMFP measurements at Carter Bridge and Estacada in September 2011 (table 9) were higher than the storm-water samples collected during the October 10, 2010, storm. These measurements suggest DOM with a high content of THM-forming carbon entered the river at that time, although water-column chlorophyll-*a* levels were not that high: 1 µg/L or less in the mainstem and less than or equal to 2.5 µg/L in North Fork Reservoir. While these specific DBP values were elevated, the overall carbon concentrations were low at this time, which resulted in relatively low absolute DBPFP values (fig. 14).

Table 9. Top ten highest carbon-normalized (specific) disinfection by-product formation potentials for filtered and unfiltered samples.

[Sampling site locations are shown in figure 1. Specific disinfection by-product formation potential concentrations calculated by dividing the filtered/unfiltered formation potentials by the dissolved organic carbon (DOC) and total organic carbon (TOC) concentrations, respectively. **Abbreviations:** THM4/HAA5, total trihalomethanes/total haloacetic acids]

Sampling site	Date	Concentration (milligrams of DBP per milligram of carbon)
Specific trihalomethane formation potential		
Filtered		
North Fork Reservoir (release depth)	09-21-11	0.208
Clackamas River at Carver	09-09-11	0.176
Clackamas River at Estacada	09-22-11	0.169
Lake Oswego–Source water	09-22-11	0.169
Clackamas River at Estacada	09-09-11	0.160
Clackamas River at Barton Bridge	09-22-11	0.158
Clackamas River Water–Source water	09-22-11	0.156
Clackamas River at Barton Bridge	09-09-11	0.149
Clackamas River at Carver	09-22-11	0.142
Clackamas River at Carter Bridge	09-08-11	0.137
Unfiltered		
Lake Oswego–Source water	09-22-11	0.201
Clackamas River at Barton Bridge	09-22-11	0.196
Clackamas River at Carver	09-22-11	0.194
Clackamas River at Barton Bridge	09-09-11	0.189
Clackamas River Water–Source water	09-22-11	0.176
Clackamas River Water–Source water	09-09-11	0.146
Clackamas River at Estacada	09-22-11	0.140
Clackamas River at Carver	09-09-11	0.135
Eagle Creek	09-08-11	0.131
Clackamas River at Estacada	09-09-11	0.126
Specific haloacetic acid formation potential		
Filtered		
Deep Creek	10-10-10	0.089
Eagle Creek	10-10-10	0.081
Clear Creek	10-10-10	0.081
Rock Creek	10-10-10	0.070
Sieben Creek	10-10-10	0.070
Sieben Creek	09-18-10	0.068
North Fork Reservoir (hypolimnion)	08-04-10	0.049
Clackamas River at Carter Bridge	10-10-10	0.049
North Fork Reservoir (metalimnion)	09-03-10	0.049
North Fork Reservoir (release depth)	08-04-10	0.048
Unfiltered		
Eagle Creek	10-10-10	0.094
Clear Creek	10-10-10	0.080
Deep Creek	10-10-10	0.069
Clackamas River Water–Source water	04-19-11	0.067
Sieben Creek	10-10-10	0.066
Rock Creek	10-10-10	0.065
Clackamas River at Carter Bridge	10-10-10	0.062
Clackamas River Water–Source water	10-10-10	0.060
Lake Oswego–Source water	10-10-10	0.055
Clear Creek	09-08-11	0.053

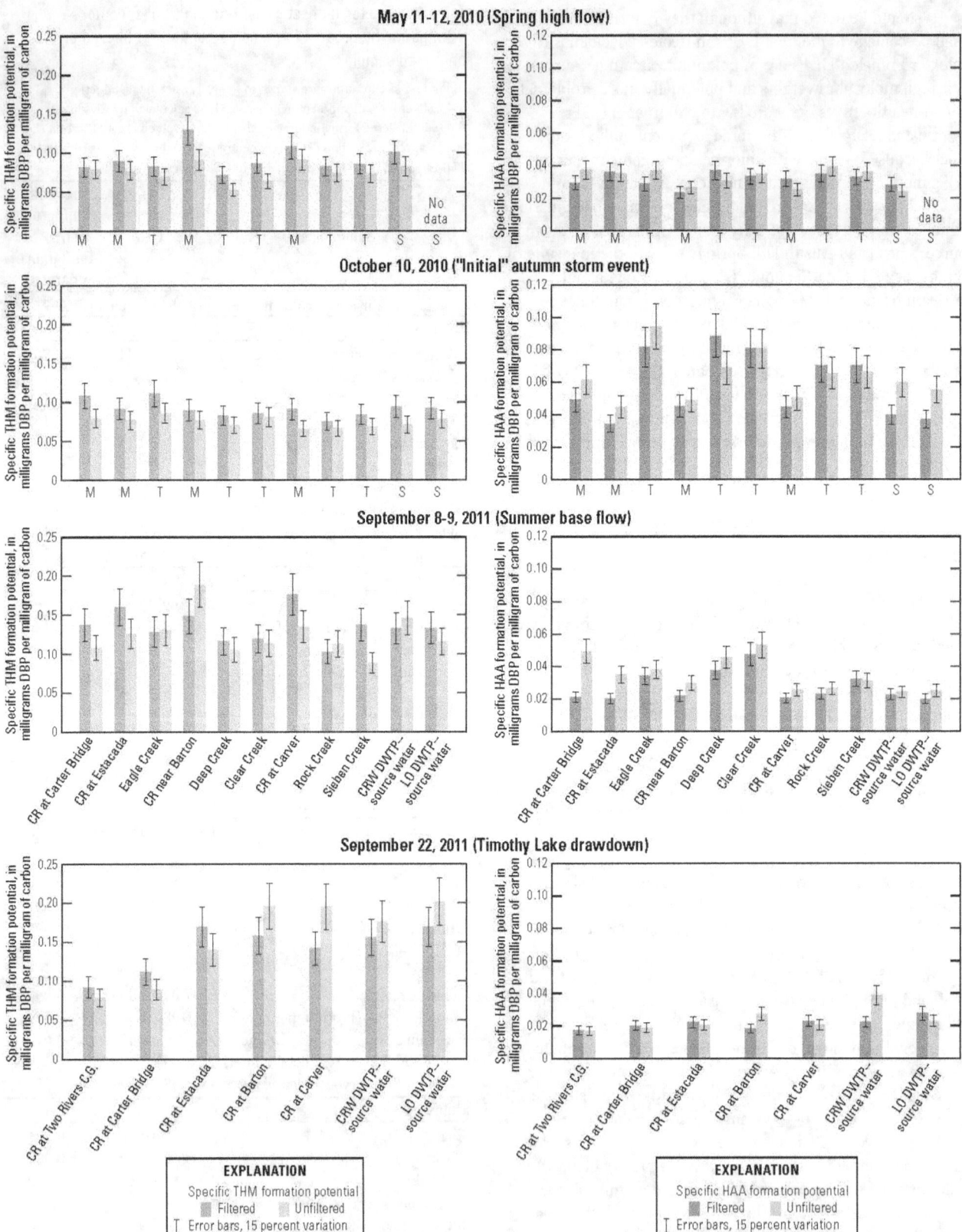

Figure 16. Specific disinfection by-product formation potential for total trihalomethanes (THM4) and total haloacetic acids (HAA5) for samples collected in the Clackamas River basin, Oregon, 2010–11. (Sites are listed in downstream order. Note variable y-axis scales. X-axis labels for Summer base flow also apply to the Spring high flow and Initial autumn storm event. Abbreviations: M, main-stem site; T, tributary; S, source-water; mg/L, milligrams per liter; CR, Clackamas River; CRW, Clackamas River Water DWTP; LO, City of Lake Oswego DWTP; DWTP, drinking-water treatment plant.)

Patterns in the STHMFP measurements for unfiltered samples were similar to filtered samples (fig. 16). The values were highest in samples collected September 2011 from main-stem and source-water sites (table 9), in samples from lower-basin tributaries (Eagle, Clear, and Rock Creeks), and from the 80-ft release depth within North Fork Reservoir.

The highest SHAAFP measurements for filtered samples were from Deep, Eagle, and Clear Creeks during the initial October 10, 2010 storm. Three samples collected from North Fork Reservoir (mid-depth, release point, and near the bottom) also had relatively high SHAAFP values during an algae bloom in August–September 2010. As with STHMFP, patterns in SHAAFP were similar for unfiltered and filtered samples; the values were highest in lower-basin tributaries during the October 2010 storm. At this time, SHAAFP values were also high in the mainstem from Carter Bridge downstream to the CRW and LO DWTP intakes. Prior studies found that HAA precursors can be linked to soil-derived, degraded DOM as well as to DOM more recently added by algal production (Kraus and others, 2008, 2011).

The correlations between STHMFP and SHAAFP were weak, which again emphasize that sources of these two classes of DBPs may differ substantially in the Clackamas Basin. Prior studies have suggested that aliphatic structures play a more important role in THM formation and aromatic structures play a more important role in HAA formation (Croué and others, 2000; Liang and Singer, 2003). A strong positive correlation was found between SHAAFP and SUVA, which suggests HAA precursors are associated with ultraviolet absorbing, aromatic compounds. This type of DOM is strongly associated with soil-derived, humified organic matter.

Higher HAA5 formation per unit carbon seen at downstream tributaries during the initial October 2010 storm, when DOC concentrations and SUVA values were elevated, agree with prior studies in this region linking HAA precursors to the flushing of organic material from soils (Kraus and others, 2010). Higher THM4 formation per unit carbon seen in the main-stem samples during September and October when SUVA values were low suggest a potential link between THM precursors and contributions of DOM from algae.

Loads and Yields of Organic Carbon and Disinfection By-Product Precursors

The highest measured concentrations of DOC and TOC were from the tributaries (fig. 9), especially Deep, Rock, and Sieben Creeks, but the limited amount of streamflow from all these tributaries resulted in relatively low DBP loads to the main-stem Clackamas River during three of four samplings (fig. 17). During storm periods, however, loads from the tributaries can become important. During the October 2010 storm, for example, the loads of DOC and TPC from Deep Creek alone accounted for about 50 and 70 percent, respectively, of that measured at the CRW DWTP intake (table 10). This carbon reacted to produce DBPs, accounting for 56 and 65 percent of the unfiltered THMFP and HAAFP loads, respectively, at the CRW DWTP intake. At this time, about half (47 percent) the total carbon was in particulate form, including duckweed fragments (see photograph 6) possibly associated with over-spillage from ponds within the Deep Creek Basin. Other streams including Eagle, Clear, and Rock Creeks contributed 28, 9, and 9 percent of the DOC load at the CRW DWTP intake, respectively, during the initial autumn storm. Eagle Creek contributed 28 and 37 percent of the unfiltered THMFP and HAAFP load, respectively, at the CRW DWTP intake (fig. 17).

THMFP in the lower mainstem for unfiltered samples increased 44 percent between Estacada and Barton on all four dates, including times when the tributaries were seemingly unimportant. This suggests a particulate source—such as sloughed benthic algae, for example—might be contributing to higher downstream THMFP. Although Eagle Creek did produce higher THMFP values compared with the mainstem at Estacada, concentrations were even higher downstream at Barton (fig. 14).

The carbon and DBP precursor yields also varied according to hydrologic condition. Forested areas with relatively higher amount of flow (upper basin, Eagle Creek, and main-stem Clackamas River downstream to Barton) had the highest DOC yields during the May 2010 spring high-flow period; Sieben, Rock, and Deep Creeks had the highest yields during the October 2010 storm (fig. 18). During the summer base-flow period, however, the lowest DBP yields were from the tributaries, and the highest yields were from the upper and middle basin. The THMFP for unfiltered samples increased from Carter Bridge downstream to Estacada and to Barton at this time (fig. 14), which could be from both phytoplankton within North Fork Reservoir and sloughed benthic algae in the reach upstream from Barton.

Photograph 6. Deep Creek running turbid during the initial autumn storm (inset shows presence of duckweed, a possible indicator of overflowing ponds in the basin). (Photographs by Kurt Carpenter, U.S. Geological Survey, October 10, 2010.)

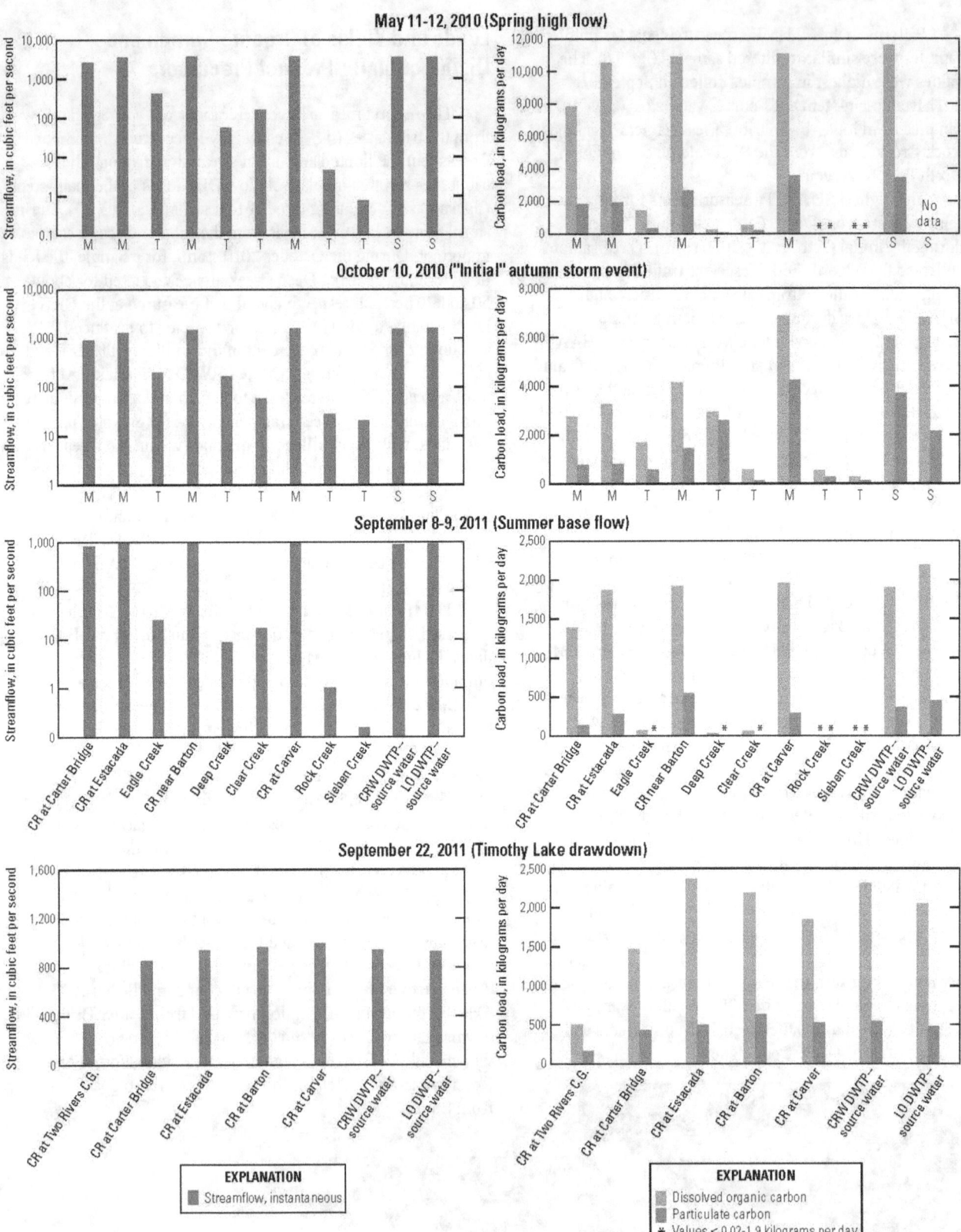

Figure 17. Streamflow and instantaneous loads of dissolved organic carbon, particulate carbon, and associated disinfection by-product formation potentials for total trihalomethanes (THM4) and total haloacetic acids (HAA5) for samples collected in the Clackamas River basin, Oregon, 2010–11. (Sites are listed in downstream order. Note log scale for y-axis in streamflow plots. X-axis labels for Summer base flow also apply to the Spring high flow and Initial autumn storm event. Loads, in kilograms per day were calculated by multiplying the concentration in mg/L by flow (x 2.447 for units conversion). Abbreviations: M, main-stem site; T, tributary; S, source-water; mg/L, milligrams per liter; CR, Clackamas River; CRW, Clackamas River Water DWTP; LO, City of Lake Oswego DWTP; DWTP, drinking-water treatment plant.)

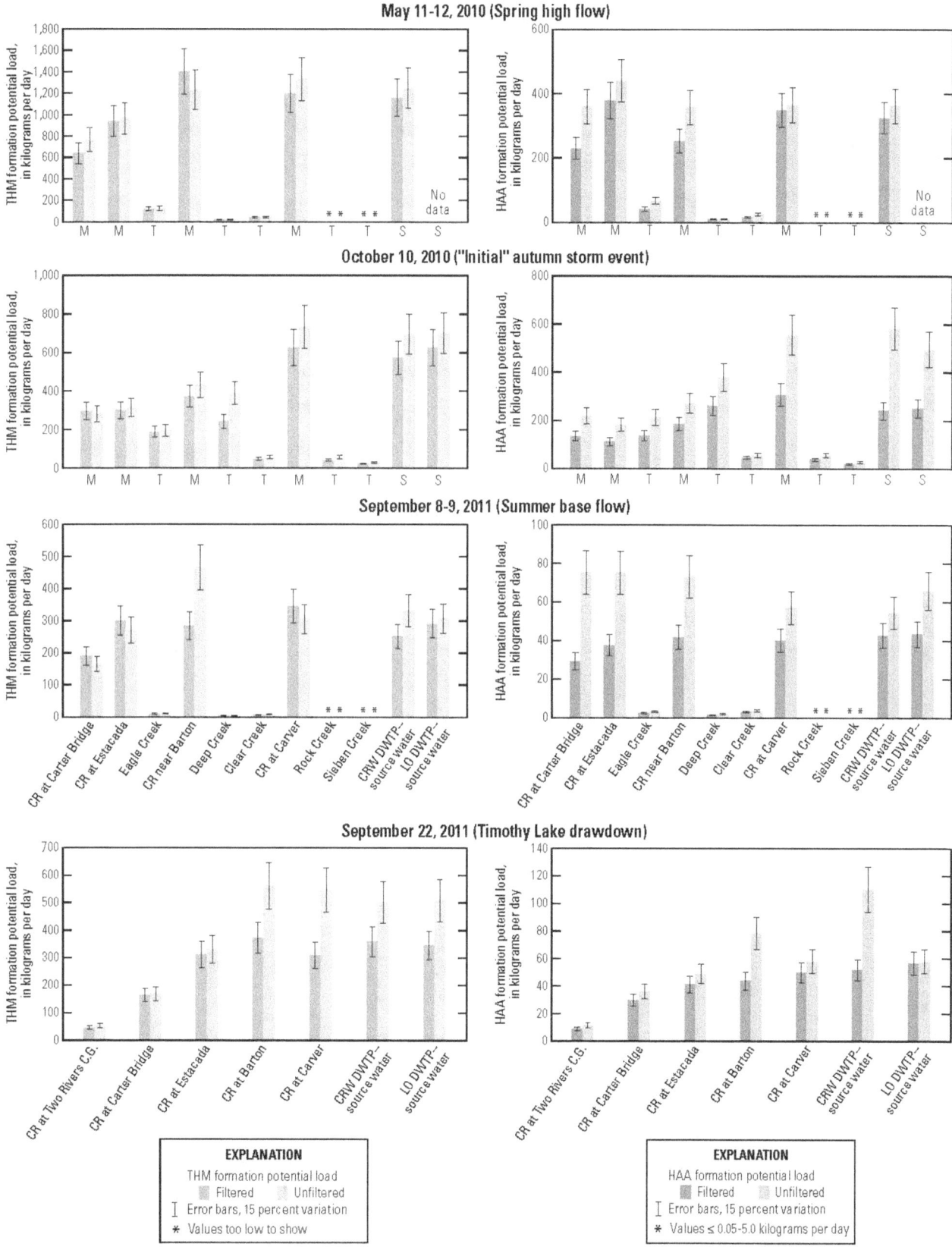

Figure 17.—Continued

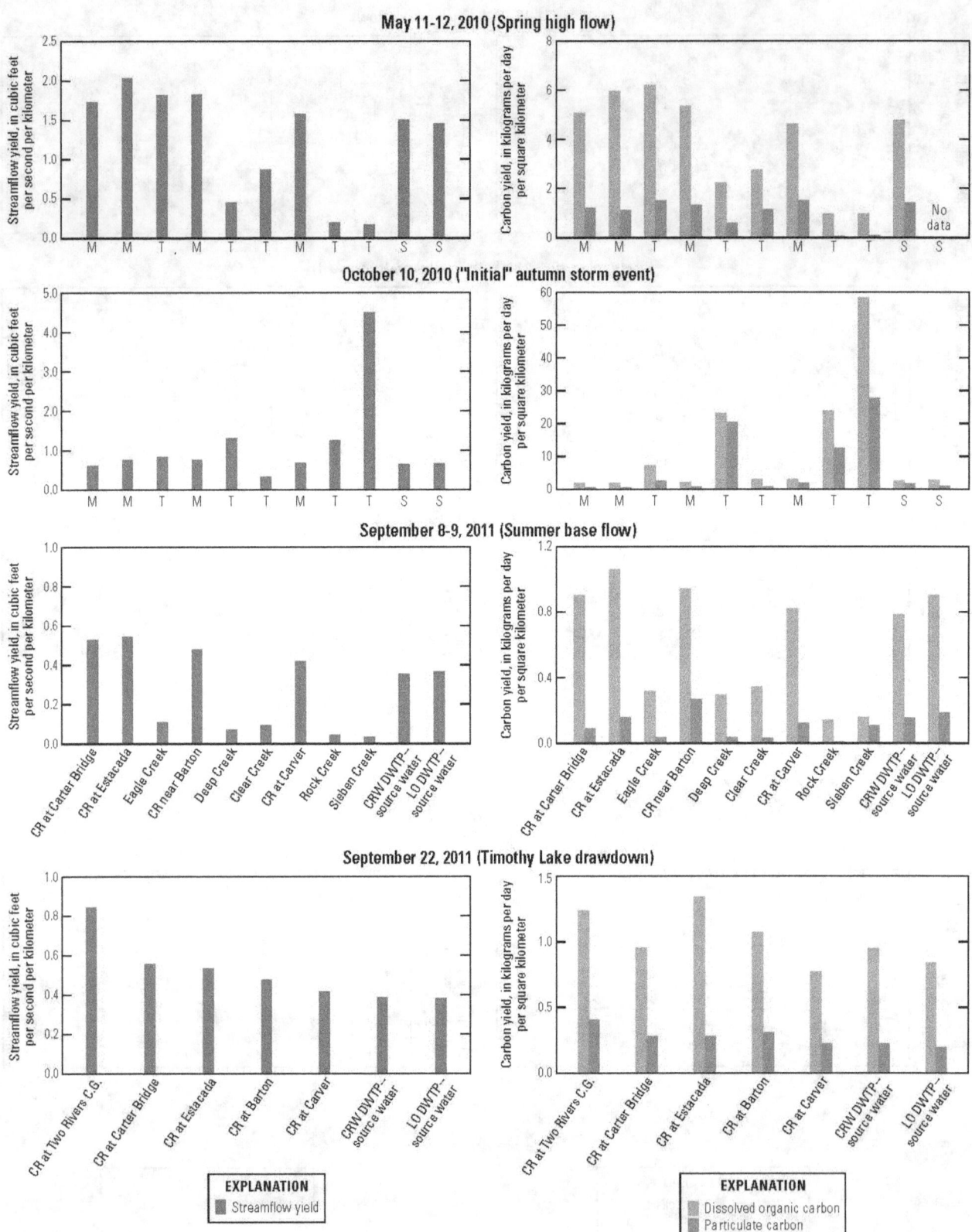

Figure 18. Instantaneous yields of streamflow, dissolved organic carbon, particulate carbon, and associated disinfection by-product formation potentials for total trihalomethanes (THM4) and total haloacetic acids (HAA5) for samples collected in the Clackamas River basin, Oregon, 2010–11. (Sites are listed in downstream order. Note the log scale for y-axis in streamflow plots. Loads, in kilograms per day were calculated by multiplying the concentration in milligrams per liter (mg/L) by flow (x 2.447 for units conversion). Yields were derived by dividing the load by the area, in square kilometers. Abbreviations: M, main-stem site; T, tributary; S, source water intake; CR, Clackamas River; CRW, Clackamas River Water DWTP; LO, City of Lake Oswego DWTP; DWTP, drinking water treatment plant.)

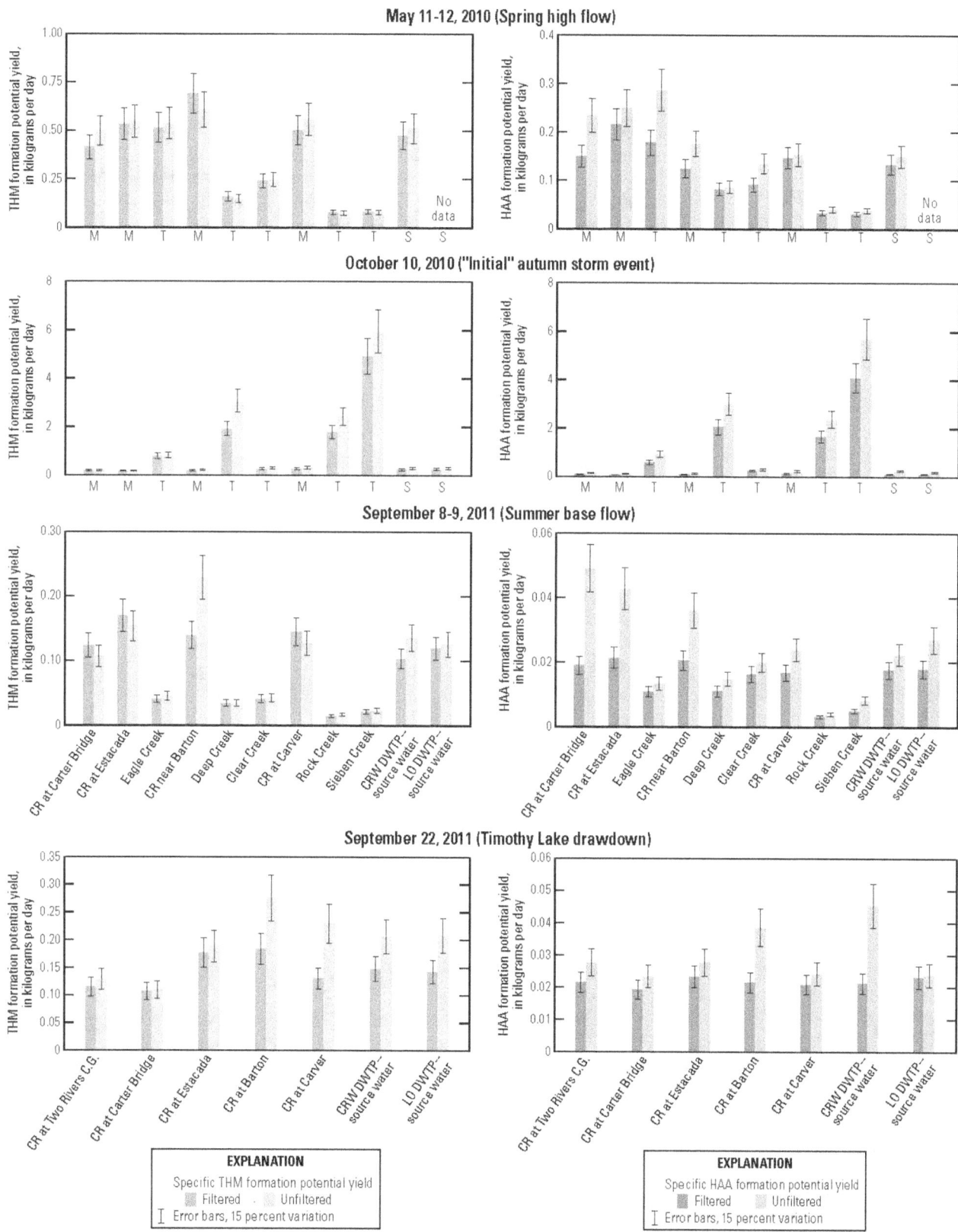

Figure 18.—Continued

Table 10. Percentage of dissolved organic carbon, total particulate carbon, and disinfection by-product formation potentials for individual sites relative to instantaneous loads at the Clackamas River Water drinking-water treatment plant intake during basin-wide sampling events, Clackamas River basin, Oregon.

[Sampling site locations are shown in figure 1. Streamflow in cubic feet per second. Abbreviations: DWTP, drinking-water treatment plant; DOC, dissolved organic carbon; TPC, total particulate carbon; THM, trihalomethane; HAA, haloacetic acid; FP, formation potential; F, filtered; U, unfiltered; –, no data]

Sampling site	River mile	Date	Streamflow	Percent of load at the Clackamas River Water DWTP intake					
				DOC	TPC	THMFP-F	THMFP-U	HAAFP-F	HAAFP-U
May 11–13, 2010 (Spring high flow)									
Clackamas River at Carter Bridge	40.8	05-11-10	2,665	67	54	55	61	71	99
Clackamas River at Estacada	23.1	05-11-10	3,579	91	57	81	77	117	121
Eagle Creek	16.7	05-11-10	422	12	10	10	10	13	18
Clackamas River at Barton Bridge	13.4	05-12-10	3,700	94	78	121	99	78	99
Deep Creek	12.1	05-11-10	58	2.4	2.2	1.7	1.5	3.2	3.0
Clear Creek	8	05-11-10	162	4.4	6.1	3.8	3.7	5.2	6.9
Clackamas River at Carver	7.9	05-12-10	3,760	95	104	103	107	108	101
Rock Creek	6.4	05-11-10	4.6	0.2	0.05	0.16	0.14	0.24	0.26
Sieben Creek	5.8	05-11-10	0.8	0.04	0.01	0.03	0.03	0.05	0.05
CRW DWTP–Source water	3.1	05-12-10	3,646	100	100	100	100	100	100
LO DWTP–Source water	0.9	05-13-10	3,549	–	–	–	–	–	–
October 10, 2010 ("Initial" autumn storm event)									
Clackamas River at Carter Bridge	40.8	10-10-10	938	45	22	52	41	56	38
Clackamas River at Estacada	23.1	10-10-10	1,338	54	22	52	45	47	31
Eagle Creek	16.7	10-10-10	197	28	16	33	28	57	37
Clackamas River at Barton Bridge	13.4	10-10-10	1,535	68	39	65	62	77	47
Deep Creek	12.1	10-10-10	167	49	70	42	56	108	65
Clear Creek	8	10-10-10	60	9	3.2	8.6	8	19	10
Clackamas River at Carver	7.9	10-10-10	1,650	113	114	109	105	127	95
Rock Creek	6.4	10-10-10	29	9	8	7.3	8.1	16	9.5
Sieben Creek	5.8	10-10-10	21	4.5	3.5	4.0	4.0	7.9	4.5
CRW DWTP–Source water	3.1	10-10-10	1,548	100	100	100	100	100	100
LO DWTP–Source water	0.9	10-10-10	1,639	112	58	109	101	104	85
September 8–9, 2011 (Summer base flow)									
Clackamas River at Carter Bridge	40.8	09-08-11	811	73	38	76	50	68	138
Clackamas River at Estacada	23.1	09-09-11	957	98	76	119	82	88	138
Eagle Creek	16.7	09-08-11	25	4	2	4	3	6	6
Clackamas River at Barton Bridge	13.4	09-09-11	980	101	148	113	140	98	134
Deep Creek	12.1	09-08-11	9	2.0	1.2	1.7	1.3	3.3	3.4
Clear Creek	8	09-08-11	18	3.4	1.5	3.0	2.4	7.1	6.8
Clackamas River at Carver	7.9	09-09-11	1,000	103	78	137	92	94	105
Rock Creek	6.4	09-08-11	1.0	0.2	0.05	0.13	0.12	0.18	0.17
Sieben Creek	5.8	09-08-11	0.2	0.0	0.14	0.04	0.03	0.05	0.07
CRW DWTP–Source water	3.1	09-09-11	864	100	100	100	100	100	100
LO DWTP–Source water	0.9	09-09-11	897	115	123	116	92	102	121
September 22, 2011 (Timothy Lake drawdown)									
Clackamas River at Two Rivers campground	57	09-16-11	343	22	30	13	10	17	10
Clackamas River at Carter Bridge	40.8	09-16-11	858	64	79	46	34	57	33
Clackamas River at Estacada	23.1	09-22-11	943	103	92	87	66	79	44
Clackamas River at Barton Bridge	13.4	09-22-11	969	95	115	104	112	84	71
Clackamas River at Carver	7.9	09-22-11	994	80	97	86	109	96	53
CRW DWTP–Source water	3.1	09-22-11	944	100	100	100	100	100	100
LO DWTP–Source water	0.9	09-22-11	930	89	88	96	101	109	52

Carbon Characterization Using Optical Properties

Optical measurements, including UVA_{254} and FDOM (ex 370/em 460 nm), were strongly correlated with concentrations of DOC (table 11) and TOC in the Clackamas River. Although DOC concentration remained relatively low in the mainstem, there were large changes in SUVA values (1.5 to 4.4 L/mg-m) over the study period, indicating a shift in DOM composition (fig. 19). SUVA values were highest during high-flow periods, indicative of contributions of higher aromatic carbon associated with humic-like material. During low-flow periods, SUVA values decreased, suggesting the DOM pool contained a greater proportion of non-aromatic, lower molecular-weight carbon derived from less-degraded organic material (Weishaar and others, 2003). SUVA increased slightly at the downstream sites in the mainstem. SUVA values were also higher in the tributary sites, particularly during the October 2010 storm event (3.2–4.3 L/g-m). Like SUVA, spectral slope values are also used to indicate changes in DOM composition. As was seen with SUVA, seasonal changes in spectral slope suggest the DOM contains higher molecular-weight, aromatic material during storm events compared to base-flow periods (appendix G1).

Contour plots of selected EEMs obtained from the fluorescence analysis are shown in fig. 20 to highlight observed changes in DOM amount and composition. The highest intensity EEMs were from Deep, Rock, and Sieben Creeks where DOC concentrations were highest. The first set of EEMs for the upper, middle, and lower mainstem during low streamflow contain a Peak T-like carbon signal near 270–340 ex/em associated with algal-derived organic matter high in protein and (or) freshly leached plant material high in polyphenols (Coble, 2007; Hernes and others, 2009; Beggs and others, 2011). The presence of this peak is best demonstrated in the last EEM from North Fork Reservoir that represents the algal grab sample collected at the surface during the September 2011 phytoplankton bloom. It should be noted that the Peak T signal was only apparent, however, in the contour plots when overall fluorescence intensities were lower, reflecting periods when DOC concentration was low. This likely results because the Peak T region is over-shadowed by the stronger humic-like signal (for example, Peaks A and C) that appeared during high-flow events (fig. 20). In order to more accurately quantify the presence of different pools of organic matter contributing to the overall EEMs spectra, PARAFAC modeling was used; this approach enables the more subtle underlying signals that are not necessarily visible in the EEM color contours to be detected and the relative contribution of each of the underlying fluorophores to be quantified.

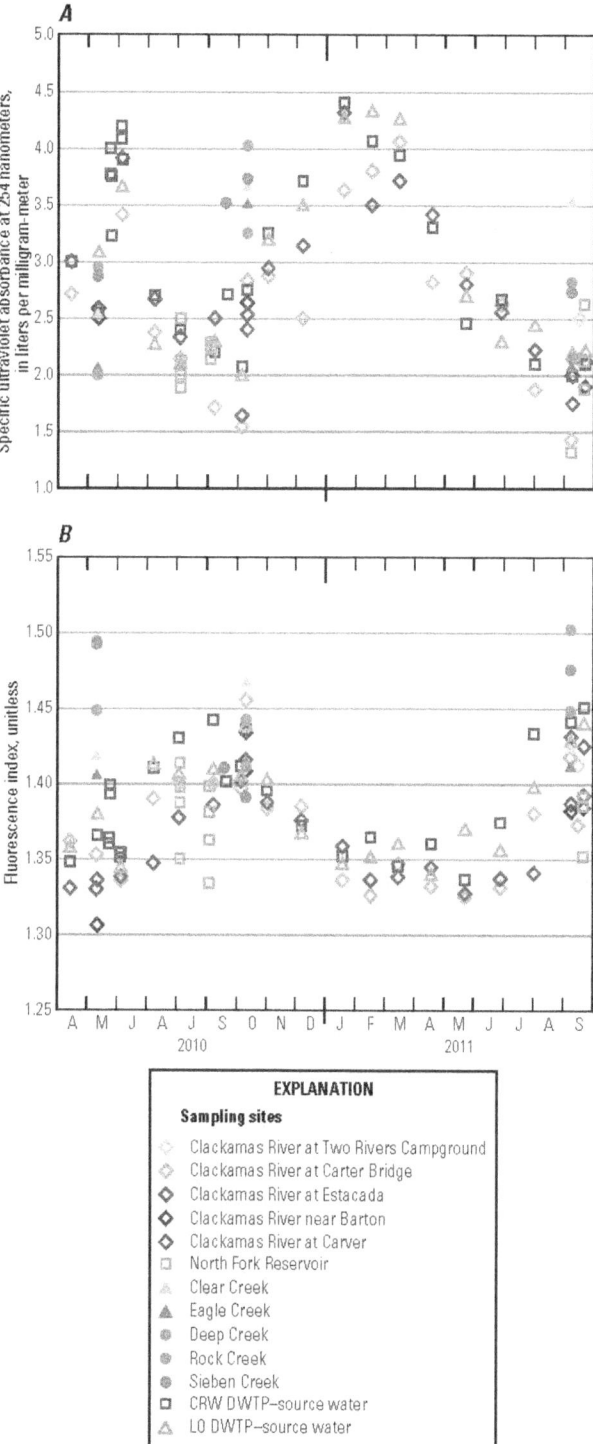

Figure 19. Seasonal patterns in (*A*) specific ultraviolet absorbance and (*B*) fluorescence index values for tributary, reservoir, and main-stem sites in the Clackamas River basin, Oregon, 2010–11. (Abbreviations: CRW, Clackamas River Water DWTP; LO, City of Lake Oswego DWTP; DWTP, drinking water treatment plant.)

Table 11. Spearman rank correlations between carbon concentrations, optical properties, and disinfection by-product concentrations in finished water and DBP formation potentials for filtered and unfiltered samples from the Clackamas River basin, Oregon, 2010–11.

[Spearman rank correlations (rho values), with P values indicated by asterisks: * (P<0.05); ** (P<0.01); *** (P<0.001). See table 6 for constituent abbreviations and units. Abbreviations: DOC, dissolved organic carbon; THM4, total trihalomethanes; HAA5, total haloacetic acids; DBP, disinfection by-product; STHMFP, carbon-specific total trihalomethane formation potential; SHAAFP, carbon-specific total haloacetic acid formation poential; F, filtered; U, unfiltered; —, not applicable or no data]

Constituent	All sites DOC	Finished water THM4	Finished water HAA5	THMFP-F	THMFP-U	HAAFP-F	HAAFP-U	STHMFP-F	STHMFP-U	SHAAFP-F	SHAAFP-U
Quantitative constituents											
Streamflow	—	**0.47	***0.64	—	—	—	***-0.51	0.27	*0.29	***-0.52	***-0.46
pH	***-0.46	-0.33	**-0.50	-0.41	**-0.40	***-0.50	***-0.51	*-0.29	-0.24	***0.71	***0.51
Turbidity	***0.72	***0.46	***0.55	***0.88	***0.93	***0.94	***0.95	*-0.29	*-0.31	***0.60	***0.47
Chlorophyll-a	***0.53	-0.20	-0.16	***0.71	***0.71	***0.70	***0.72	*-0.32	-0.28	***0.75	***0.58
Total organic carbon	***0.98	***0.63	***0.71	***0.95	***0.96	***0.98	***0.98	*-0.34	-0.28	***0.77	***0.62
Dissolved organic carbon	—	***0.72	***0.83	***0.96	***0.95	***0.98	***0.97	-0.28	-0.26	***0.69	***0.49
Total particulate carbon	***0.87	0.32	***0.71	***0.87	***0.93	***0.93	***0.94	-0.33	-0.27	***0.78	***0.60
UVA254	***0.96	***0.67	***0.79	***0.94	***0.96	***0.98	***0.94	-0.34	*-0.28	***0.73	***0.58
Dissolved organic matter fluorescence	***0.98	***0.68	***0.81	***0.95	***0.93	***0.95	***0.91	**-0.34	-0.28	***0.69	***0.53
PEAK A	***0.96	***0.65	***0.77	***0.93	***0.91	***0.92	***0.93	**-0.34	-0.28	***0.72	***0.56
PEAK C	***0.98	***0.67	***0.80	***0.94	***0.92	***0.94	***0.92	**-0.34	-0.28	***0.69	***0.53
PEAK M	***0.96	***0.64	***0.71	***0.93	***0.91	***0.92	***0.95	**-0.34	-0.28	***0.74	***0.59
PEAK D	***0.98	***0.68	***0.81	***0.96	***0.93	***0.95	***0.93	**-0.37	-0.30	***0.72	***0.51
PEAK B	***0.89	***0.60	***0.54	***0.85	***0.90	***0.93	***0.92	**-0.39	-0.32	***0.69	***0.51
PEAK T	***0.92	**0.46	**0.41	***0.88	***0.90	***0.92	***0.91	**-0.35	-0.29	***0.68	***0.51
PEAK N	***0.94	***0.53	***0.53	***0.91	***0.91	***0.92	***0.87	*-0.33	-0.27	***0.63	***0.47
C1 Loading	***0.91	***0.51	***0.54	***0.88	***0.87	***0.87	***0.95	*-0.34	*-0.28	***0.74	***0.59
C2 Loading	***0.98	***0.67	***0.80	***0.96	***0.94	***0.96	***0.95	*-0.34	*-0.28	***0.72	***0.56
C3 Loading	***0.97	***0.68	***0.81	***0.95	***0.93	***0.94	***0.93	*-0.34	*-0.29	***0.73	***0.58
C4 Loading	***0.98	***0.68	***0.80	***0.95	***0.93	***0.95	***0.94	*-0.34	*-0.29	***0.69	***0.58
C5 Loading	***0.89	**0.50	**0.44	***0.85	***0.90	***0.92	***0.92	**-0.37	*-0.29	***0.69	***0.49
Qualitative constituents											
SUVA	***0.49	***0.44	***0.59	***0.72	***0.67	***0.73	***0.73	**-0.39	**-0.37	***0.78	***0.70
Spectral slope (S275–295)	***-0.46	**-0.40	-0.27	***-0.45	**-0.42	***-0.45	***-0.45	0.23	0.12	***-0.52	***-0.56
Spectral slope (S290–350)	-0.18	-0.30	-0.19	-0.18	-0.23	-0.26	-0.26	-0.04	-0.10	-0.26	-0.28
Spectral slope (S350–400)	0.24	0.05	0.23	0.26	0.23	0.19	0.18	-0.03	0.04	0.18	0.09
Spectral slope ratio	***-0.43	-0.30	**-0.37	**-0.44	-0.41	***-0.4	**-0.39	0.19	0.08	**-0.45	***-0.39
Fluorescence Index	0.18	-0.25	**-0.44	*0.28	0.23	0.17	0.17	-0.01	0.09	0.13	0.20
Humic Index	0.07	-0.12	-0.03	0.21	0.10	0.04	0.03	0.10	0.08	0.07	0.15
Percent component 1	***0.55	-0.04	-0.11	***0.57	***0.52	***0.48	***0.48	-0.23	-0.19	*0.29	***0.28
Percent component 2	-0.02	*0.37	***0.55	-0.03	-0.09	-0.08	-0.08	0.16	0.07	0.03	0.15
Percent component 3	-0.16	0.02	0.23	-0.14	-0.18	-0.24	-0.25	***0.38	0.32	-0.18	-0.10
Percent component 4	-0.23	0.03	0.20	-0.16	-0.23	-0.24	-0.25	***0.36	0.26	-0.06	0.04
Percent component 5	-0.24	-0.21	***-0.45	***-0.41	***-0.30	-0.24	-0.23	-0.10	-0.06	-0.18	-0.27

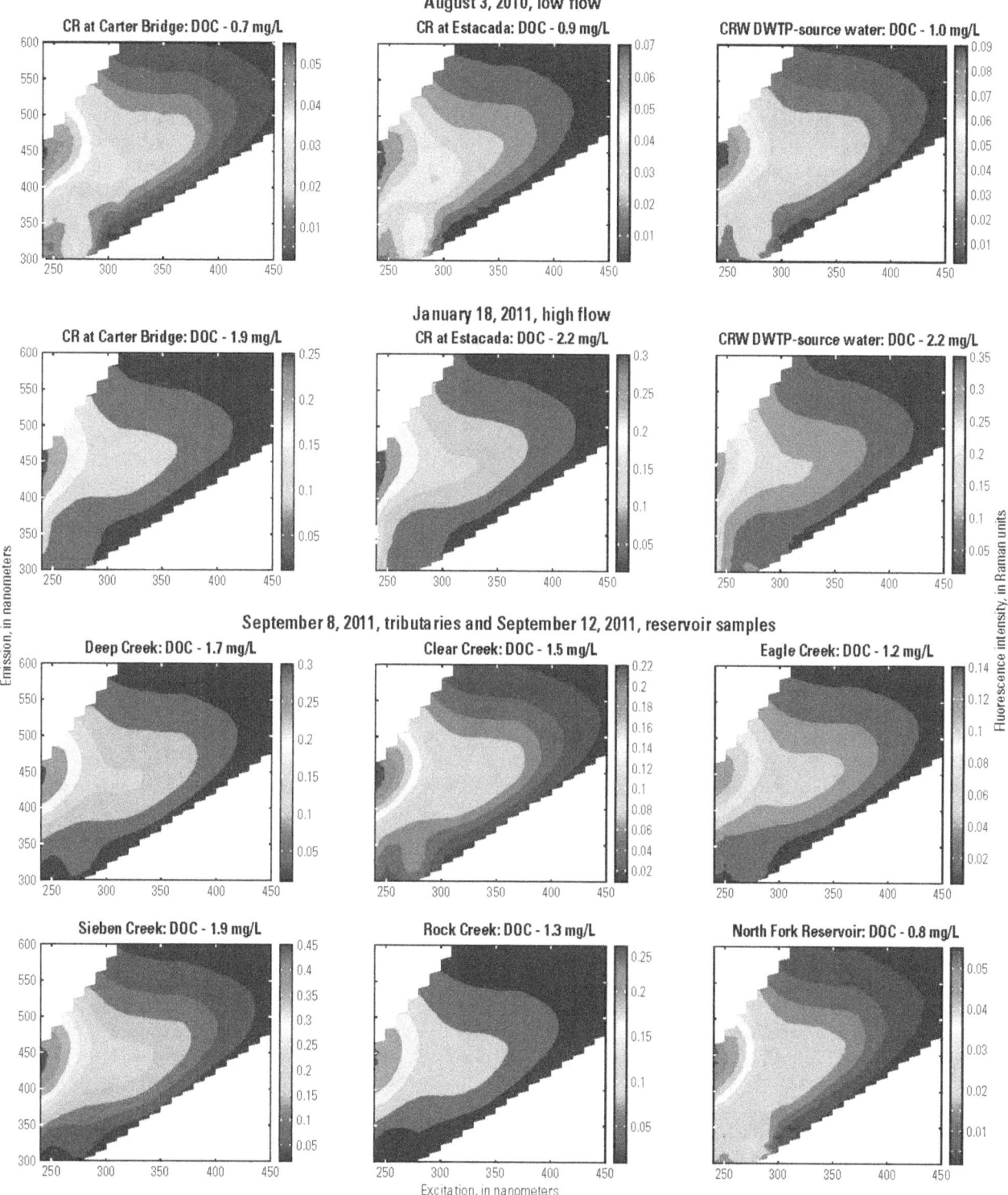

Figure 20.   Excitation–emission matrices (EEMs) for water samples from selected main-stem, tributary, and reservoir sites across a range of stream conditions, Clackamas River basin, Oregon, 2010–11. (Abbreviations: DOC, dissolved organic carbon; CR, Clackamas River; CRW, Clackamas River Water drinking-water treatment plant; mg/L, milligrams per liter.)

The PARAFAC model developed and validated with 167 samples from the mainstem, tributary, North Fork Reservoir, and source and finished water produced five components: C1–C5 (fig. 21). Model components C2, C3, and C4 are associated with terrestrial humic-like substances commonly associated with fluorescence Peaks A and C (Stedmon and Markager, 2005; Coble, 2007; Murphy and others, 2008; Yamashita and others, 2008). There are, however, subtle differences among these three components. Fluorescence associated with C1 has been associated with several different sources of DOM, including marine and terrestrial humic acids that have been microbially processed, freshly produced (phytoplankton derived, for example) labile material identified as Peak N, and material exported from agricultural and wastewater-impacted catchments identified as Peak M (Stedmon and Markager, 2005; Coble, 2007; Fellman and others, 2010). C2 is typical of terrestrial organic matter composed of high molecular-weight and aromatic compounds (McKnight and others, 2001; Stedmon and others, 2003). C5, located in a region frequently referred to as "protein-like" because tryptophan and tyrosine fluoresce in this region, is associated with less-processed carbon derived from fresh terrestrial plants, algae, and (or) wastewater (Murphy and others, 2008; Hernes and others, 2009; Beggs and others, 2011).

Examination of the relative contributions of the different PARAFAC components shows that C1, C3, and C4 represent the bulk of DOM fluorescence (fig. 22). Component C5 generally represented a smaller and more highly variable fraction. Although there were slight seasonal shifts in the relative proportions of the different PARAFAC components in CRW source water (fig. 22B), the overall trend shows a consistent pattern. Average percentages were 26 percent for C3, 21–22 percent for C1 and C4, and 15–16 percent for C2 and C5, reflecting dominance by terrestrial types of carbon in the lower mainstem.

The fluorescence spectra of Eagle and Clear Creeks, two streams draining predominantly forested basins (table 2), were similar to the main-stem sites (fig. 22), reflecting dominance by terrestrial humic substances. The other three tributaries— Deep, Rock, and Sieben Creeks—have less forested area and are variously influenced by agriculture and urban development (table 2). These streams contained the highest proportion of component C1 (24–39 percent; fig. 22), suggesting the DOM contributed from these tributaries differed in composition compared to the other sites. Rock and Sieben Creeks also showed the lowest proportion of C5.

The fluorescence spectra of samples from North Fork Reservoir had the highest average percentage of component C5 (fig. 22), likely reflecting the presence of recently added DOM from phytoplankton algae that were prevalent in the reservoir during all samplings. Organic matter recently contributed by algae is expected to contain a greater protein-like signal.

Although there were only minor changes in the FI values (1.3–1.5 across the watershed samples), seasonal trends suggest DOM in the mainstem is more dominated by microbial-derived carbon between August and October (fig. 19), when more algal contributions would be expected. This trend is in accordance with the SUVA values. FI values were also higher in the tributary sites compared to main-stem sites for the May 2010 and September 2011 basin-wide samplings. The FI value was highest (2.3) for the algal grab sample collected from the surface of North Fork Reservoir in September 2011 (not shown in fig. 19), which contained a high abundance of blue-green algae (*Anabaena* sp.). Although the FI did not vary a lot, these results suggest this measure could be indicative of algal-derived carbon despite the relatively low levels of chlorophyll-*a* observed during the study.

The HIX (table 1) ranged from about 2 to 8; the most notable trends were higher values in some of the tributaries, particularly in September 2011, possibly from highly-processed humified material, and lower values for reservoir samples, indicative of more recently added "fresh" material (appendix G1).

Carbon characterization based on optical properties provided complementary evidence to the carbon concentrations during the two autumn storms in 2010 that offers insights into the different effects of these storms on watershed processes. The second, larger storm (fig. 13), for example, produced higher SUVA and HIX values and lower FI values generally associated with greater aromatic content and higher molecular-weight material. This was accompanied by declines in the percentages of carbon components C1 and C5 and increases in components C2, C3, and C4—all indicative of the carbon quality shifting from a more labile microbial/algal source with lower aromaticity during the initial storm to a higher molecular-weight, aromatic source of terrestrial origin after the larger storm (table 8). These results point to the success of fluorescence technologies in detecting these shifts in DOM quality (and quantity) and provide a means by which the quality of source water can be closely monitored for treatment-plant operations, for gaining a deeper understanding of river conditions and processes, and for evaluating trends over short and long time scales. Through implementation of these instruments in studies like this, new technologies are being developed to advance these capabilities.

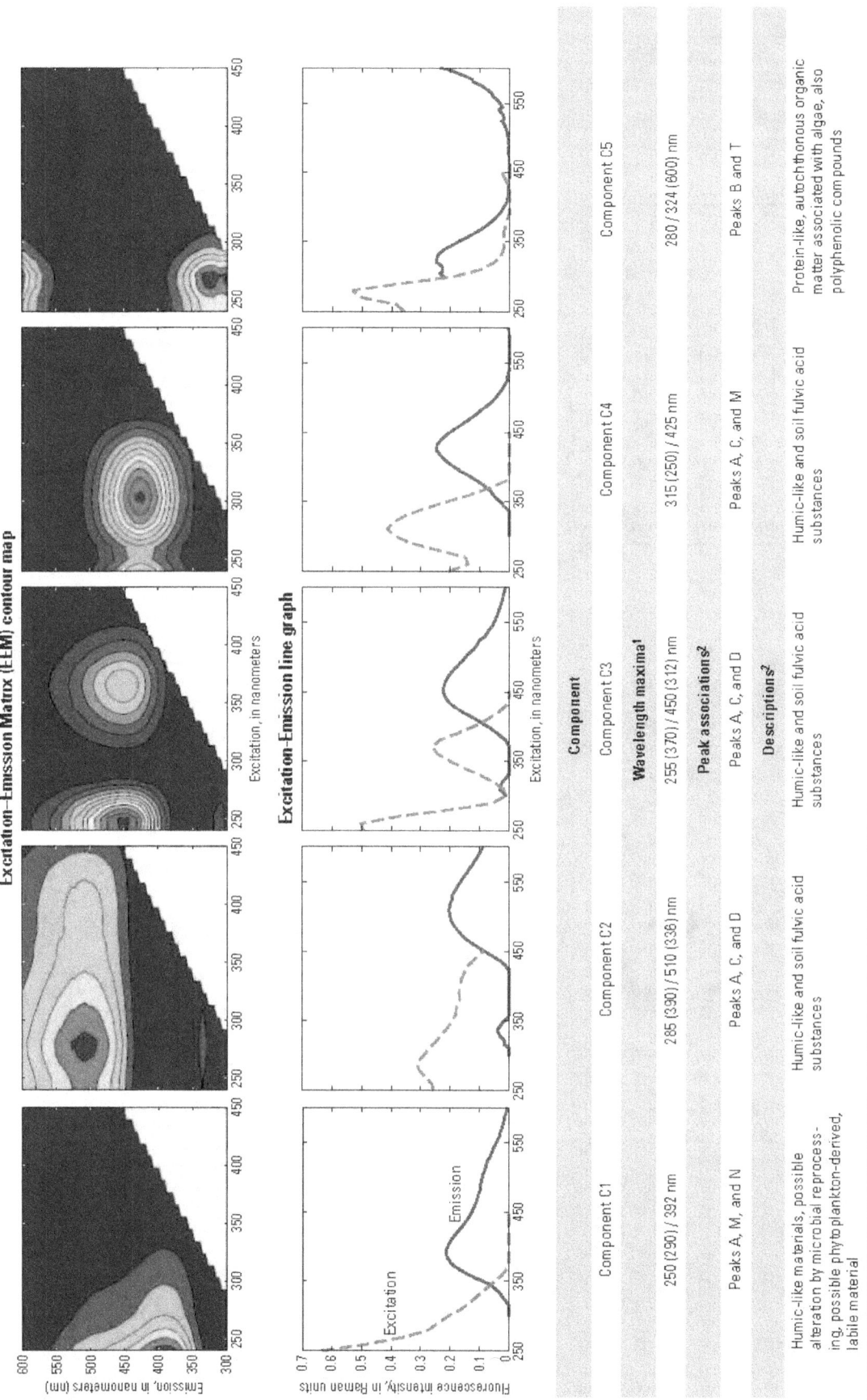

Figure 21. Principal fluorescence components identified in the PARAFAC model for the Clackamas River basin, Oregon, 2010–11.

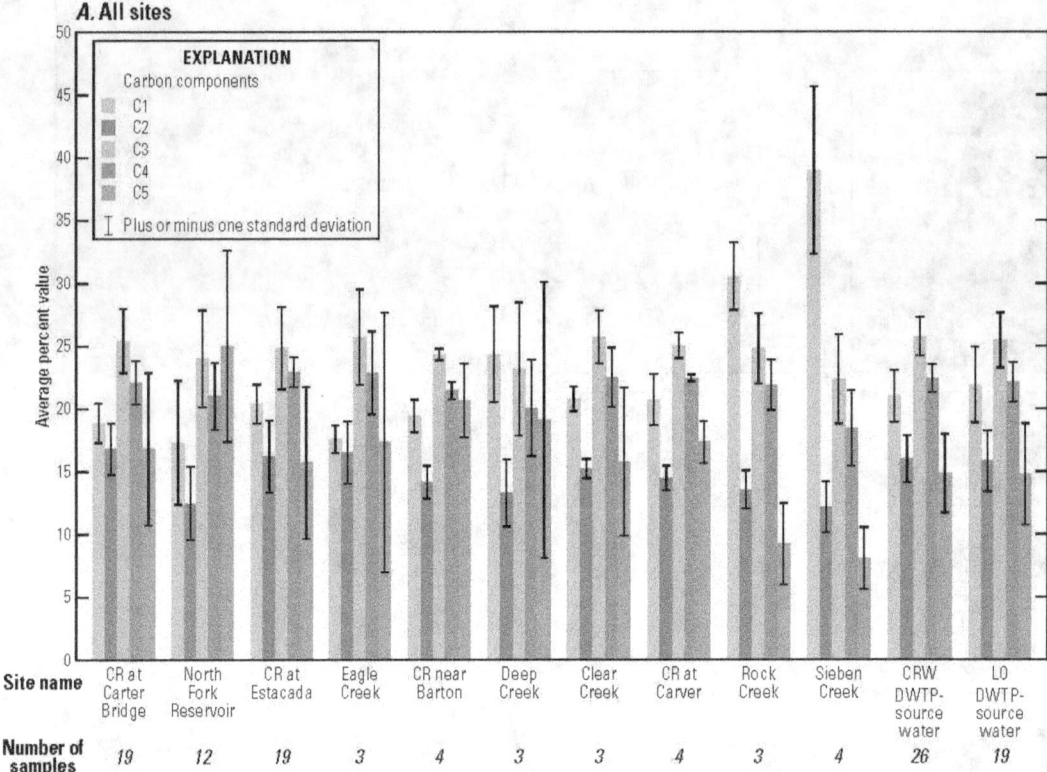

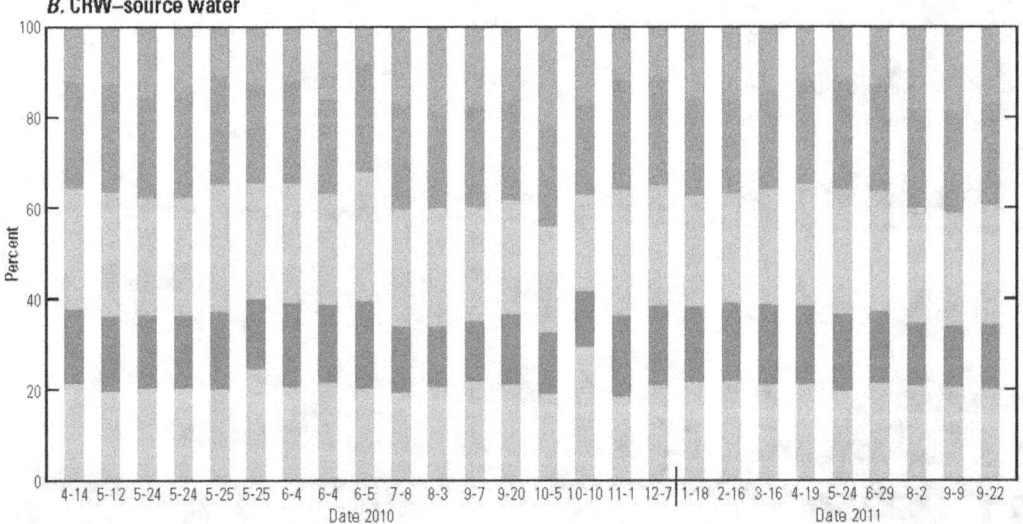

Figure 22. Relative percentages of PARAFAC components in water samples collected from (*A*) all sites, and (*B*) source water at the Clackamas River Water drinking-water treatment plant, Clackamas River basin, Oregon, 2010–11.

Proxies for Carbon and Disinfection By-Product Precursor Concentrations

With growing concerns about the effects of DBPs on human health and associated changes in the regulatory standards, there is much interest in identifying the sources of DBP precursors and developing tools to better monitor and predict carbon and DBP concentrations. To identify source-water attributes that are good predictors or "surrogates" for DBP formation, relations between concentrations of THM and HAA (in finished water and DBPFP samples) and organic carbon concentrations and optical properties were examined using Spearman rank correlations (table 11). While many of the variables were significantly positively correlated with concentrations of THM4 and HAA5 in finished water as well as with THMFP and HAAFP, the best DBP predictors were DOC concentration, FDOM and several fluorescence peaks, and total component loadings from components C2, C3, and C4. In general, the correlations between concentrations of TOC and DOC and DBPs in drinking-water samples were somewhat higher for HAA5 compared with THM4, although all were significant ($p < 0.001$).

Laboratory DBPFPs were highly correlated with concentrations of DOC and TOC, turbidity, UVA, fluorescence peak intensities, PARAFAC component loadings, and various other indicators ($r > 0.9$, $p < 0.001$; table 11); r values for THMFP and HAAFP were similar. Correlations between STHMFP, filtered and unfiltered, and other constituents were not that high ($r < 0.4$), although some correlations were significant (table 11). In contrast, SHAAFP was highly correlated with a host of other indicators; for example, SHAAFP in filtered-water was significantly correlated ($r = 0.78$, $p < 0.001$) with SUVA (table 11).

Given that DBP precursors are strongly correlated with DOC concentrations, there is much interest in identifying DOC proxies, especially for water systems where DBPs are approaching drinking-water standards. Laboratory bench-top measurements of FDOM were highly correlated ($r = 0.98$; $p < 0.001$) with DOC concentration (fig. 23) as well as laboratory THMFP and HAAFP (table 11). In addition, continuous in-situ FDOM measurements were highly correlated with laboratory FDOM ($r = 0.99$; $p < 0.001$; appendix F2) and provided similarly high correlations with DOC concentration ($r = 0.96$; $p < 0.001$) as did the laboratory based optical measurements (fig. 23; appendix F2). These strong correlations provide convincing evidence that in-situ FDOM measurements can be effective at tracking concentrations of DOC and DBP precursors in source water. In addition to providing high-frequency, real-time data, the in-situ measurements do not require the sample collection, processing, and analyses needed for UVA measurement.

The ability to monitor DWTP inflow water quality to predict finished-water DBPFP continuously, in real-time, would be of great benefit for DWTP operations. In addition to source-water quality, however, the amount of DBPs that form during treatment is also influenced by DWTP operations, including coagulation, disinfection type and dose, contact times, pH, temperature, and other factors. Much can be learned by using this information along with monitoring feedback to adaptively manage treatment plants and optimize for DBP precursor removal.

During this study, finished-water DBPs were determined approximately monthly, resulting in 18 data points to compare in-situ FDOM measurements to finished-water THM4 and HAA5 concentrations. The correlation between in-situ FDOM and finished-water HAA5 concentration was significant ($r = 0.83$; $p < 0.001$; appendix F3); applying the equation derived from this relation, the high-frequency in-situ FDOM data were used to estimate finished-water HAA5 concentrations (fig. 24). If accurate, these estimates would suggest HAA5 concentrations did not exceed the 0.06 mg/L MCL during this study. While the correlation between in-situ FDOM and chloroform was significant ($r = 0.64$; $p < 0.01$), the relation between FDOM and finished-water THM4 was not, so estimates of continuous finished-water THM concentrations were not generated for this study. Further research on this topic is warranted given that fluorescence was a good predictor of laboratory THM formation potential, here and elsewhere (Hua and others, 2007; Marhaba and others, 2009; Kraus and others, 2010). It should also be emphasized that the relation between source-water quality and DBP formation potentials conducted on untreated water in the laboratory under uniform conditions is expected to be stronger than the correlation with treated (finished) water that has undergone coagulation (table 11).

The significant, positive correlation between SHAAFP and SUVA (table 11) suggests the HAA precursor pool is made up of, if not linked to, chromophoric DOM, which may explain why there is a stronger correlation between FDOM and HAA5 concentrations compared to THM4. Prior studies indicate HAA precursors are more aromatic compared to THM precursors (Croué and others, 2000; Liang and Singer, 2003; Hong and others, 2008), and HAA precursors have been associated with terrestrially derived fulvic and humic acids (Kraus and others, 2008, 2010). However, other studies have also observed a link between HAA precursors and algal-derived DOM (Chen and others, 2008; Hong and others, 2008; Kraus and others, 2011).

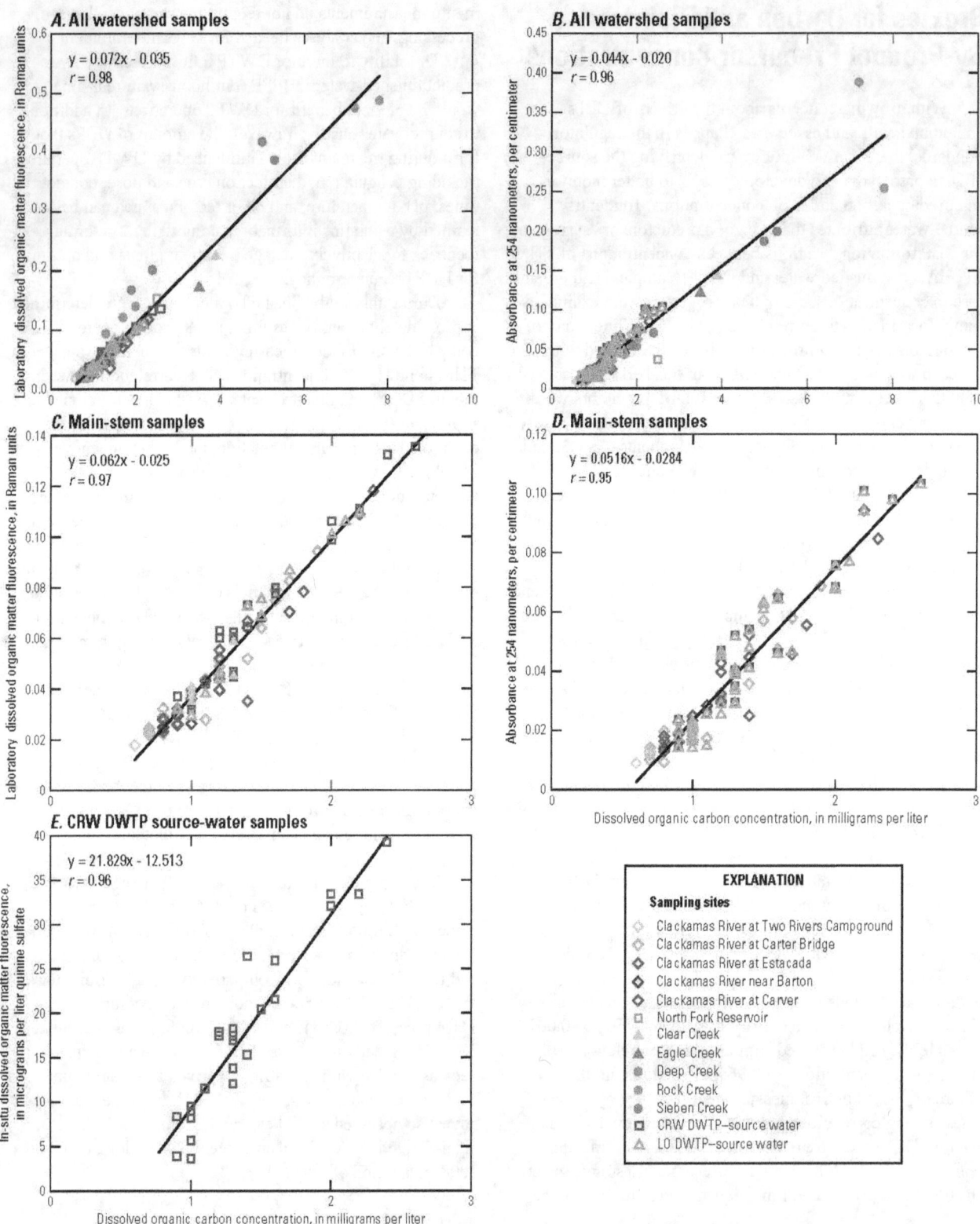

Figure 23. Relation between concentrations of dissolved organic carbon and (*A*) laboratory measurement of fluorescent dissolved organic matter (FDOM), (*B*) laboratory measurement of UVA$_{254}$, (*C*) laboratory measurement of FDOM for main-stem samples, (*D*) laboratory measurement of UVA$_{254}$, and (*E*) in-situ FDOM, from the Clackamas River basin, Oregon, 2010–11. (Note variable axes values.)

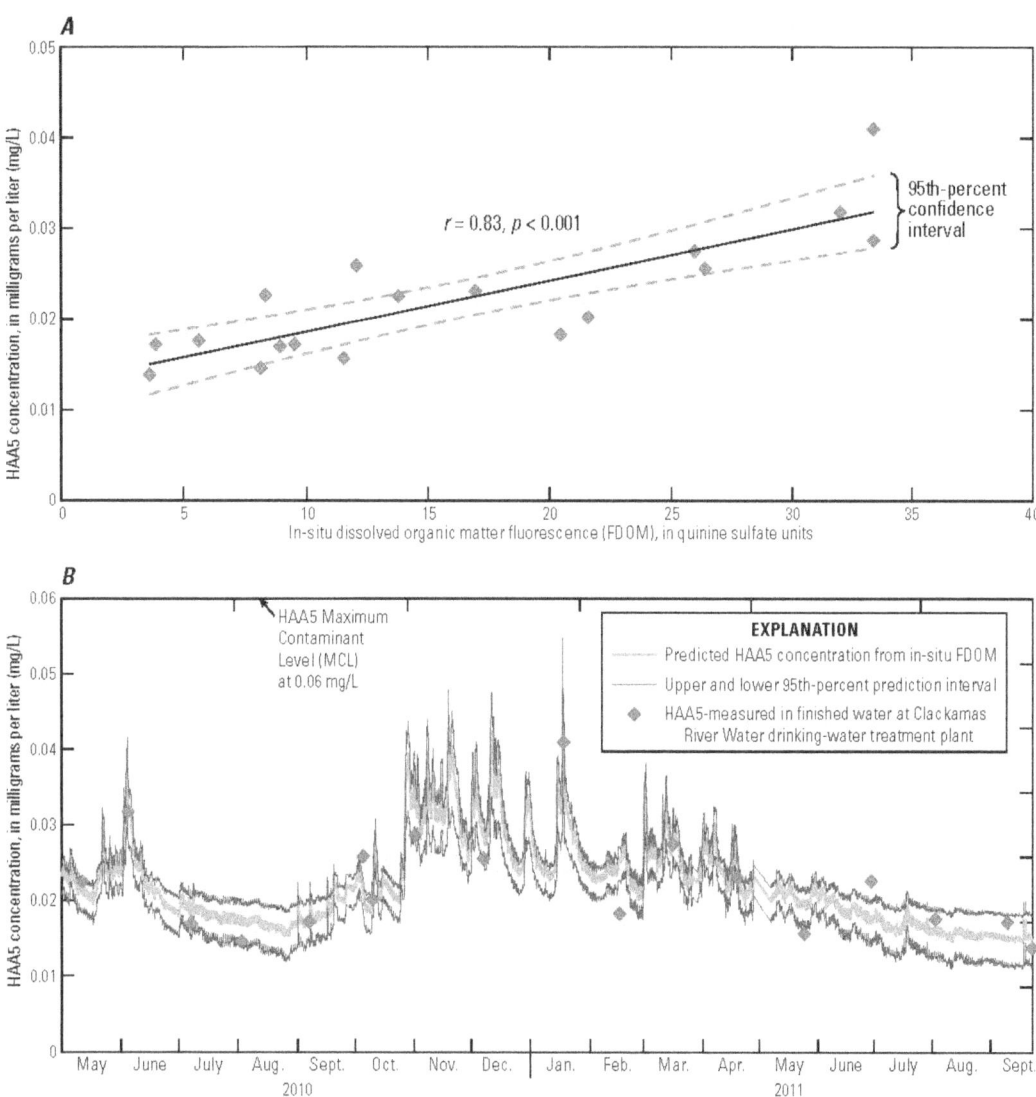

Figure 24.   (*A*) Relation between in-situ dissolved organic matter fluorescence (FDOM) and total haloacetic acid (HAA5) concentrations in finished water, and (*B*) predicted HAA5 concentrations in finished water from in-situ FDOM at the Clackamas River Water drinking-water treatment plant, 2010–11.

Water Treatment and Removal Efficiencies for Dissolved Organic Carbon and Disinfection By-Product Precursors

Water-treatment processes are designed to remove organic carbon, among other constituents, to minimize the formation of DBPs. The CRW and LO DWTPs coagulate and chlorinate raw source water simultaneously, which reduces DOC concentrations in finished water to about 0.7 mg/L (fig. 25). The loss of DOC during treatment was partially dependent on the source-water DOC concentration entering the treatment plant such that when DOC concentrations were elevated, the percentage of DOC removed increased (up to about 50 percent). Although these samples with high concentrations of DOC resulted in the highest percentage of DOC removed, they also had the highest DBP concentrations in finished water (fig. 11).

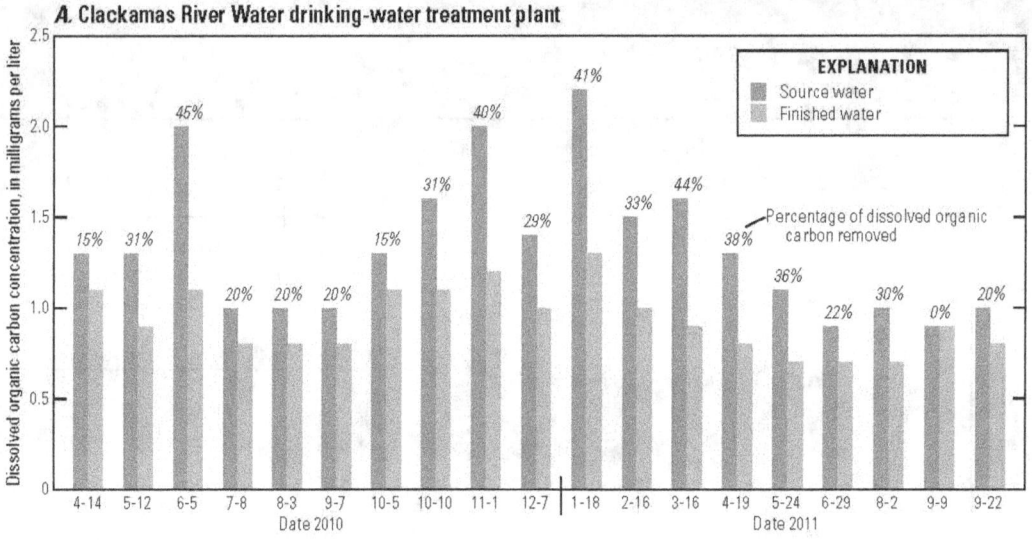

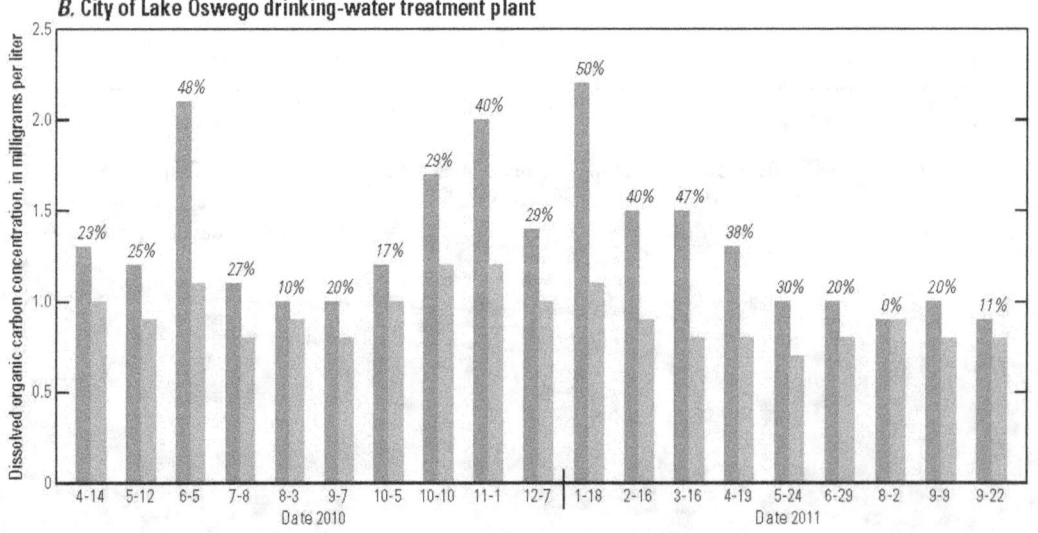

Figure 25. The seasonal pattern in source- and finished-water dissolved organic carbon concentrations and percent DOC removal at (A) Clackamas River Water and (B) City of Lake Oswego drinking-water treatment plants, Clackamas River basin, Oregon, 2010–11.

During the water treatment process, there was a 60–85 percent reduction in fluorescing material between source and finished water; 3 representative sample pairs are shown in fig. 26. There was also a marked decrease in SUVA and an increase in the FI values, suggesting the DOM remaining in finished water following coagulation and chlorination contained lower aromatic content and lower molecular-weight substances, as is commonly reported (Kitis and others, 2001; Sharp and others, 2006; Beggs and others, 2009). Compared with source water, finished water had consistently higher percentages of C1 relative to the other components, although the total fluorescence loading from C1 was lower indicating some removal (fig. 26). Because C1 was highest in samples from the tributary sites most heavily impacted by anthropogenic activities (fig. 20), there may be a connection between these land uses and the presence of DOM that is less amenable to removal by standard coagulation. Additional jar test experiments with similar fluorescence measurement might identify treatment methods that target removal of such "C1 carbon" that might lead to reduced DBP concentrations in finished water.

Results from the treatability jar-test experiments conducted on CRW DWTP source water during the four basin-wide surveys provided information on the amount of DOC and DBP precursors that could be removed by coagulation itself and in combination with PAC. Source-water DOC concentrations during the experiments ranged from 0.9 to 1.7 mg/L (fig. 27). Coagulation with optimum doses of alum and ACH coagulant reduced DOC concentrations 30 to 39 percent; the removal rate was highest when the DOC concentrations were highest during the October 2010 storm (table 12). Following coagulation, SUVA decreased and FI values increased, indicating preferential removal of the higher molecular-weight, aromatic carbon.

Co-addition of PAC and coagulants led to slightly higher removal of DOC (4–10 percent higher) and lower DBPFP values compared to coagulant alone for all jar tests except the one conducted on November 9, 2011, which showed no difference (fig. 27). These decreases, however, only represented a further reduction in DOC concentration of about 0.1 to 0.2 mg/L. The amount of DOM removed by the laboratory jar tests was higher than that measured in finished water by about 20 percent, which was equivalent to about 0.2 mg/L DOC. The reduction in DOC concentration following coagulation led to a similar decrease in THM precursor concentrations; however, the reduction in HAA precursor concentrations was much greater. HAAFPs decreased about 70 percent during all four jar tests (table 12).

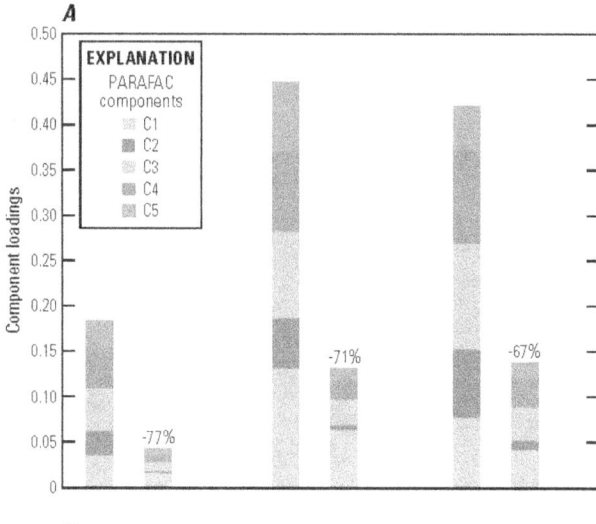

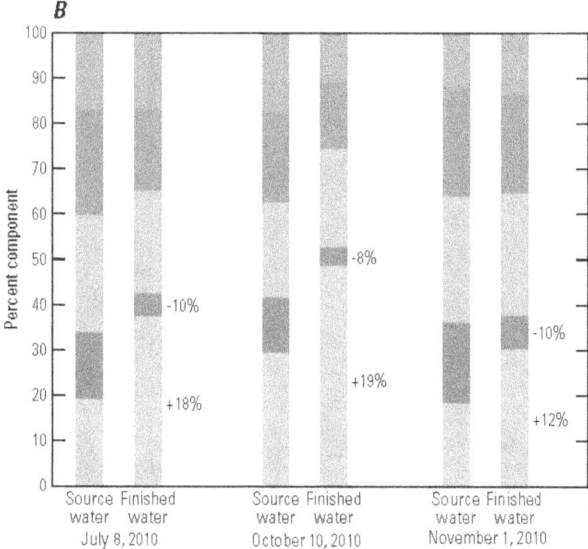

Figure 26. Effect of water treatment on (*A*) component loading, and (*B*) percentage of each component in source and finished water during July 2010 low flow, October 2010 initial storm, and November 2010 major-flush event at the Clackamas River Water drinking-water treatment plant, Clackamas River basin, Oregon. (Numbers show percent reduction in total fluorescence in (*A*), and the largest changes in selected component percentages in (*B*). See figure 21 for explanation of carbon components.)

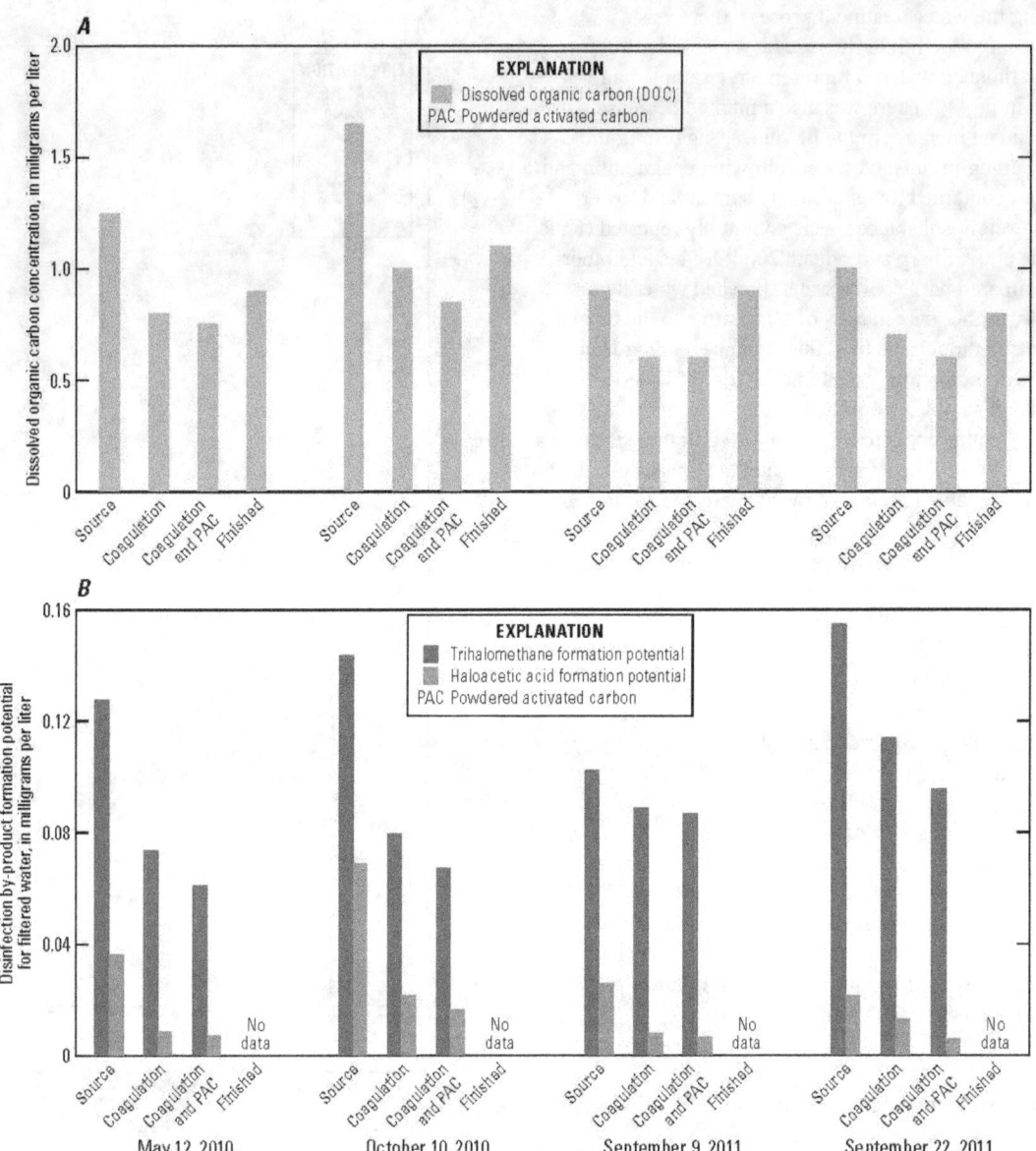

Figure 27.   Results of treatability jar-test experiments showing the effect of coagulation (aluminum sulfate and aluminum chlorhydrate) and powdered activated carbon (PAC) on concentrations of (*A*) dissolved organic carbon and (*B*) disinfection by-product precursors at the Clackamas River Water drinking-water treatment plant, Clackamas River basin, Oregon, 2010–11.

Table 12. Percent removal of dissolved organic carbon and disinfection by-product formation precursors during treatability jar-test experiments conducted at the Clackamas River Water drinking-water treatment plant, Clackamas River basin, Oregon, 2010–11.

[Average values ± one standard deviation. Removal of dissolved organic carbon (DOC) from source to finished water is shown for comparison. Abbreviations: DBP, disinfection by-product; THMs, trihalomethanes; HAAs, haloacetic acids; PAC, powdered activated carbon; –, not applicable or no data]

| Date | Treatment | Percent removal | | |
| | | DOC | DBP precursors | |
			THMs	HAAs
05-12-10	Coagulation	36 –	42 ± 5.3	76 ± 1.6
	Coagulation + PAC	40 ± 5.7	52 ± 5.4	79 ± 1.2
	Source-finished water[1]	28 –	– –	– –
10-10-10	Coagulation	39 ± 0.0	45 ± 1.8	69 –
	Coagulation + PAC	48 ± 4.3	53 ± 1.1	76 ± 2.2
	Source-finished water[1]	33 –	– –	– –
09-09-11	Coagulation	33 ± 0.0	13 ± 2.9	69 ± 1.4
	Coagulation + PAC	33 ± 0.0	15 ± 10.0	74 ± 5.4
	Source-finished water[1]	0 –	– –	– –
09-22-11	Coagulation	30 ± 0.0	27 –	65 –
	Coagulation + PAC	40 ± 0.0	38 ± 4.1	73 ± 1.0
	Source-finished water[1]	20 –	– –	– –

[1]Percent reductions in DOC during actual treatment at the Clackamas River Water drinking-water treatment plant.

The use of coagulation during drinking-water treatment removes TPC and DOC. Because DOC concentrations in the Clackamas River have been historically low, the CRW and LO DWTPs were designed primarily to target the removal of particles during treatment; thus, the coagulation and chlorination steps take place simultaneously. Removal of TPC and DOC prior to chlorination would, however, likely further reduce the formation of THMs and HAAs. Results from the jar tests showed coagulation with alum and ACH is particularly effective at removing HAA precursors. The preferential removal of HAA precursors over the bulk DOM pool and THM precursors agrees with the finding that the HAA precursor pool is associated with more aromatic, high molecular-weight material and that the pool of DOM is more amenable to removal by coagulation (Croué and others, 2000; Liang and Singer, 2003; Hong and others, 2008). Because the BQ values for HAAs in finished water are occasionally elevated, upgrades to treatment plants that allow for coagulation and filtration prior to chlorination may, therefore, be a design improvement.

The CRW and LO DWTPs occasionally use PAC during water treatment (up to 5 mg/L at CRW and up to 50 mg/L at LO) to control for tastes and odors. In contrast to studies that have found substantial removal of HAAs and THMs with commercial charcoal-filtration water pitchers (Levesque and others, 2006), the addition of PAC provided only minor improvement in DOC and DBP precursor removal, based on the four jar tests conducted here. A full evaluation of this method to reduce DBP precursors may be warranted because these data only reflect conditions during the study period. The benefits of PAC may be more pronounced during periods of higher algal contributions when the DOM pool contains higher fractions of low-SUVA, high-FI material.

Custom In-Situ Fluorescence Sensors

The three custom fluorometers designed for this study were deployed at the CRW DWTP intake, but only for a limited time, and late in the study period (fig. 28 and table 4). The majority of this period covered the prolonged recession to the summer base-flow period, through autumn, and into the January rainy season. Individually, the two sensors centered near Peak C designed to measure the FI tracked DOC concentration and were highly correlated to FDOM from the standard Cyclops-7 fluorometer, which is also focused on this Peak C region but has a wider band-pass (table 5).

While both of the Peak C custom sensors tracked DOC concentration, there was a change in the ratio of these two sensors, referred to here as $FI_{in\text{-}situ}$ (fig. 28). Prior to this study, the FI has only been calculated using benchtop measurements of fluorescence; the ratio of emission at 470 to 520 nm at an excitation wavelength of 370 nm effectively provides information about the slope of the fluorescence response in this region of EEMs space (McKnight and others, 2001). As described above, the FI provides information about DOM source and composition. The two in-situ sensors developed for this study, however, have a broader band pass than benchtop instruments. Thus, while the ratio of these two sensors is expected to be correlated to the benchtop FI ratio, the absolute value of the $FI_{in\text{-}situ}$ is expected to differ from what is commonly reported for benchtop measurements of surface-water samples—a range of 1.2 to 1.8 (Cory and others, 2010). The $FI_{in\text{-}situ}$ values for this study ranged from 0.9 to about 1.2 (fig. 28).

Over the 10-month deployment of these sensors, the $FI_{in\text{-}situ}$ values first decreased as streamflow and DOC concentration decreased and then increased in September through December 2011 when streamflow and DOC concentrations remained low and relatively stable (fig. 28).

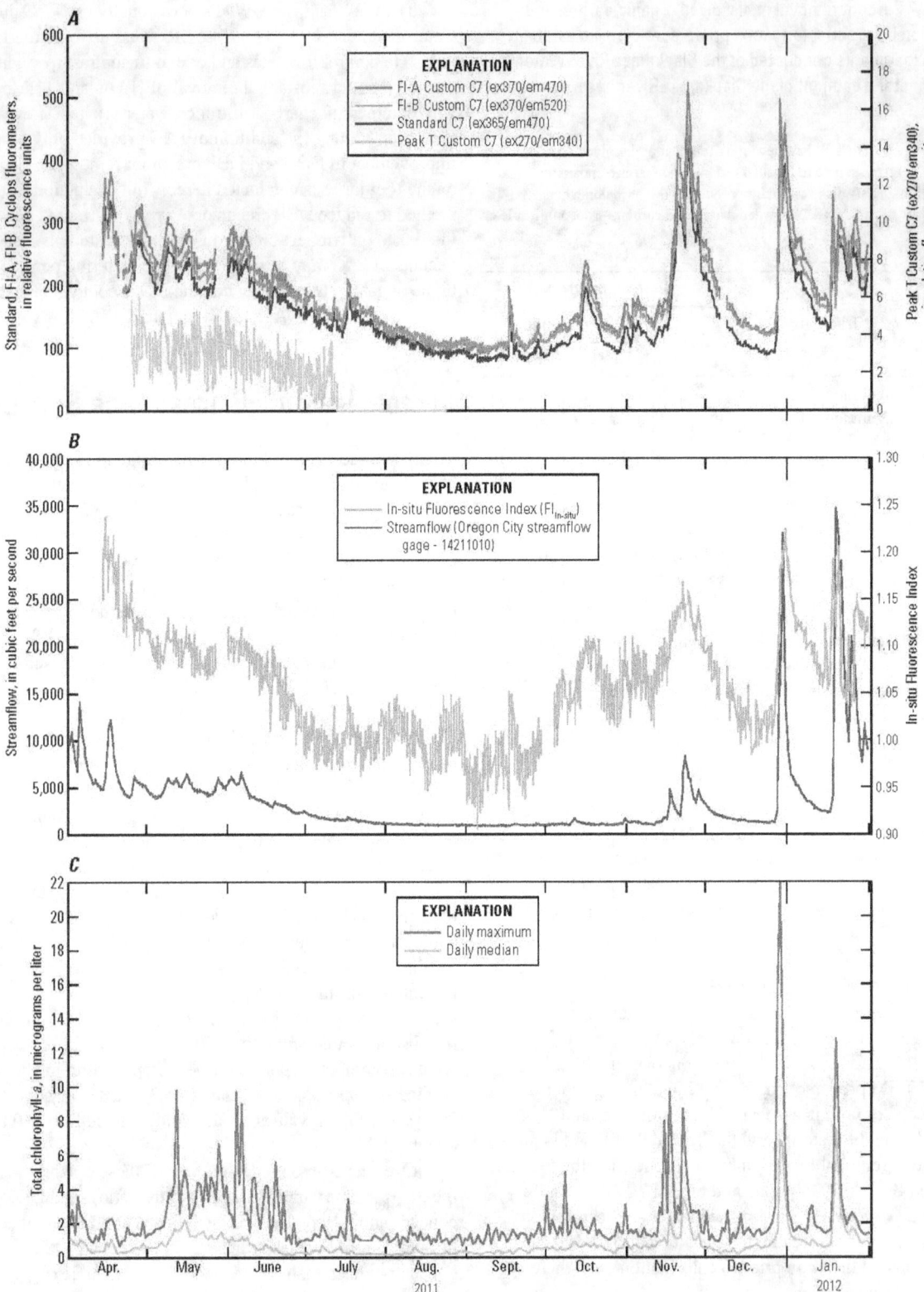

Figure 28. (*A*) Continuous fluorescence from the standard and three custom fluorescent dissolved organic matter sensors, (*B*) streamflow and in-situ Fluorescence Index, and (*C*) daily median and maximum water-column chlorophyll-*a* in the lower Clackamas River, Oregon, 2011–12.

Because FI values are assumed to be independent of concentration, these trends suggest a change in carbon composition. The initial decrease in $FI_{in-situ}$ suggests a trend towards a DOM pool increasingly dominated by terrestrial, high molecular-weight material. The increase during the later period of deployment in September–November coincided with the drawdown of Timothy Lake, and a period when benthic algal populations may have started to senesce at the end of the growing season. Concentrations of chlorophyll-*a* at the Oregon City monitor produced occasional spikes up to about 10 µg/L (fig. 28); these periodic peaks suggest moderate sloughing of benthic algae. With the onset of rain, $FI_{in-situ}$ values increased, perhaps because of suspension of decaying algae in the river or some other factor. Although data collected for this study are not conclusive, these initial results show a dynamic response in the $FI_{in-situ}$ to changes in DOM character and quantity, and show promise for future water-quality monitoring applications.

The design of the experimental Peak T custom sensor (ex 270/em 340 nm) required the use of a low-ultraviolet LED (table 5). The signal from this sensor was low, likely because of the low amount of fluorescence response from DOM in this region of the EEMs. Furthermore, the loss of signal from the sensor in mid-July was determined to be caused by loss of signal output from the lamp. Testing of LEDs in this region has shown that photon output from these deep ultraviolet LEDs are short-lived. As LED manufacturers continue to improve the efficiency of deep ultraviolet LEDs—the ability of the device to convert electrons to photons or the external quantum efficiency—these deep ultraviolet LED-based field sensors are expected to improve concurrently. Testing of these low-ultraviolet fluorometers in waters that are known to have higher fluorescence in this region (wastewater impacted rivers, for example; Hudson and others, 2008; Goldman and others, 2012), may also prove to be a more suitable application for these sensors.

Sources of Organic Carbon that Contribute Disinfection By-Product Precursors

Many lines of evidence were considered to evaluate possible sources of DOM that form regulated DBPs in the Clackamas River basin. The individual samples provided some indication as to where and when samples had elevated DBPFP values, and longitudinal and temporal variations in the data were useful for identifying patterns in the mainstem, especially when combined with Data Grapher analyses of the continuous water-quality data from the four-station monitoring network (fig. 1).

Considering that STHMFP and SHAAFP values were not correlated (appendixes F5 and F6), these two classes of DBPs seem to have different fundamental sources, which may require different watershed-management strategies for control. The filtered STHMFP value from North Fork Reservoir during an algal bloom (see photograph 2a-b) was the single highest value, which points to this source as worthy of future monitoring should THM concentrations increase in the future. The highest specific, or carbon-normalized, DBPFP values (table 9 and fig. 16) suggest that carbon most prone to producing THMs came from the reservoir and lower main-stem Clackamas River in September 2011; carbon contributing to HAA formation was highest in the three largest lower-basin tributaries, Eagle, Clear, and Deep Creeks during the initial October 2010 storm (fig. 14). These streams drain basins containing large proportions of private timberland, rural residential, and some agricultural land, primarily Christmas tree plantations, nurseries, cane berries, and some row crops (fig. 3). The greatest amount of agricultural land in the basin, about 16 mi², is contained within the Deep Creek Basin where nurseries and Christmas tree farms are abundant (Carpenter, 2003). It is, however, unclear to what degree each of these potential sources contributes DBP precursors to the mainstem and downstream drinking-water intakes.

While lower-basin tributaries had the highest concentrations of organic carbon (DOC and TOC) and DBP precursors (fig. 17), because of the relatively low flows from these streams most of the time, the primary sources of carbon are located in the upper forested basin where most of the flow also originates. This finding is consistent with a study of the nearby McKenzie River that also found DBP precursors to be primarily derived from upper-basin terrestrial carbon (Kraus and others, 2010). Our results also show elevated THM- and HAA-precursor concentrations when DOC concentrations also were elevated (fig. 11), and DBPs were formed primarily from chlorination of dissolved, terrestrially-derived organic compounds such as humic and fulvic acids. This is consistent with results for the DBPFPs that showed the dissolved fraction dominating the precursor pool (fig. 14); although particulate carbon did contribute some DBP precursors, most were dissolved.

Terrestrial Inputs

Multiple lines of evidence indicate terrestrial inputs are the dominant sources of carbon to the lower Clackamas River. First, most of the DBPFP loads in the lower Clackamas River at the CRW DWTP are already accounted for at Estacada (table 10), which drains a mostly forested basin. Second, the fluorescence data and PARAFAC model identified five dominant components, four of which are soil-related (fig. 21). Third, the co-dominant HAA in finished water was TCAA, an organic compound widely found in forest soils (McCulloch, 2002).

TCAA is a chlorinated hydrocarbon with many sources. It is used for many industrial purposes, including the synthesis of other organic compounds, and as an herbicide, for example, but it is also a breakdown product of TCE and other solvents. TCAA is also formed during various chlorination processes, in wood-pulp processing and drinking-water treatment (McCulloch, 2002). TCAA is measureable in the atmosphere at concentrations ranging from less than 0.02 µg/L in Switzerland up to 20 µg/L in urban Berlin, Germany; global average concentrations are about 0.5 µg/L (McCulloch, 2002). Although TCAA is not volatile, it is highly soluble in water and can precipitate in rain; this explains its prevalence in forest soils, especially coniferous soils (McCulloch, 2002). While it is possible there is some background level of TCAA in the Clackamas River derived through this process, this has not been investigated.

Although unstudied here, the erosion of TCAA-containing soils might be a factor in the prevalence of TCAA in drinking water from the Clackamas River. In a previous study, Carpenter (2003) found that levels of total phosphorus (TP) in the forested tributaries of the Clackamas Basin were highly correlated with percentage of "non-forest upland"—mostly timber harvest areas ($r = 0.96$, $p <0.001$). TP also was correlated with silica concentrations ($r = 0.84$, $p <0.001$), which suggests soil erosion could be a source of phosphorus to the river. It is possible the loss of particulate and dissolved carbon from the forested watershed areas are lost in a similar fashion through erosion, although leaching of DOM from organic soil horizons also certainly contributes to riverine carbon.

The flushing of decomposed organic matter by autumn and winter rains is a complex process governed by the types of organic matter and microbes present, thickness of the vadose zone and local water-table dynamics, temperature, and other factors. Decomposition of plant materials is a key process that successively transforms solid organic matter into fine particles, colloids, and DOM solutes that can leach into surface waters and form DBP precursors.

It is hypothesized here that increased deposition of nitrogen, from atmospheric and other sources, has accelerated decomposition rates of organic matter in portions of the watershed including deforested and previously harvested regrowth forests. Because of the large store of bulk carbon in the forest and the extensive timber harvesting in the basin (see numerous U.S. Forest Service "watershed analyses" reports referenced in Carpenter [2003]), decomposition of organic matter and burned slash left over from previous harvest operations might be an important source of carbon to the river that may have increased over time and may help explain the increase in DBPs observed over the last three decades (fig. 4). Although Federal forestland in the Pacific Northwest, including most of the Clackamas Basin, has been recovering since the Northwest Forest Plan was enacted in 1993, timber harvesting was extensive in the late 1980s and early 1990s, which produced a large reservoir of decaying organic matter that may have subsequently leached DOM. Previous studies cited in Turner and others (2011) suggest that estimates of forest carbon stocks in the region covered by the Northwest Forest Plan had declined substantially during 1953–87 from high rates of timber harvesting. In recent years, however, regrowth of many of these forests has increased net ecosystem productivity, thereby possibly shifting the balance toward storing carbon through sequestration (Turner and others, 2011). Whether or how this may change export of DBP precursors is, however, not known but could be examined with further study.

Algae

Even though algae fix carbon through photosynthesis and contribute to organic-matter pools, the degree to which benthic and phytoplankton algae contribute DBP precursors in the Clackamas River remains an important unanswered question. Although it is well-established that algae can be a source of DOM-containing DBP precursors, there is conflicting evidence regarding whether DOM produced by algae is more or less reactive per unit carbon compared with terrestrial sources (Jack and others, 2002; Nguyen and others, 2005; Kraus and others, 2011). Differences in DBP-precursor content of algal-derived DOM likely arise from a combination of factors, including algal species, growth stage, release of extracellular material, and environmental processing of algal-derived material and compounds (Jack and others, 2002; Nguyen and others, 2005).

To initially address this question, correlations were examined between chlorophyll-*a* and DBPFP (fig. 29). This approach focuses on the viable particulate algal cells that fluoresce and did not include non-fluorescing particles, such as dead or decaying algal filaments. Although the correlations between chlorophyll-*a* and THMFP were weak, the relation with HAAFP was significant ($p <0.001$) for filtered ($r = 0.56$) and unfiltered ($r = 0.49$) samples (fig. 29).

Other lines of evidence suggest algae are having an influence over carbon amount and quality in the Clackamas River. While water-column chlorophyll-*a* concentrations were atypically low in the lower mainstem (fig. 8), there were several observations from the data that not only demonstrate how algae can affect carbon amounts and DOM quality in the river, but also provide some evidence algae is contributing THM precursors.

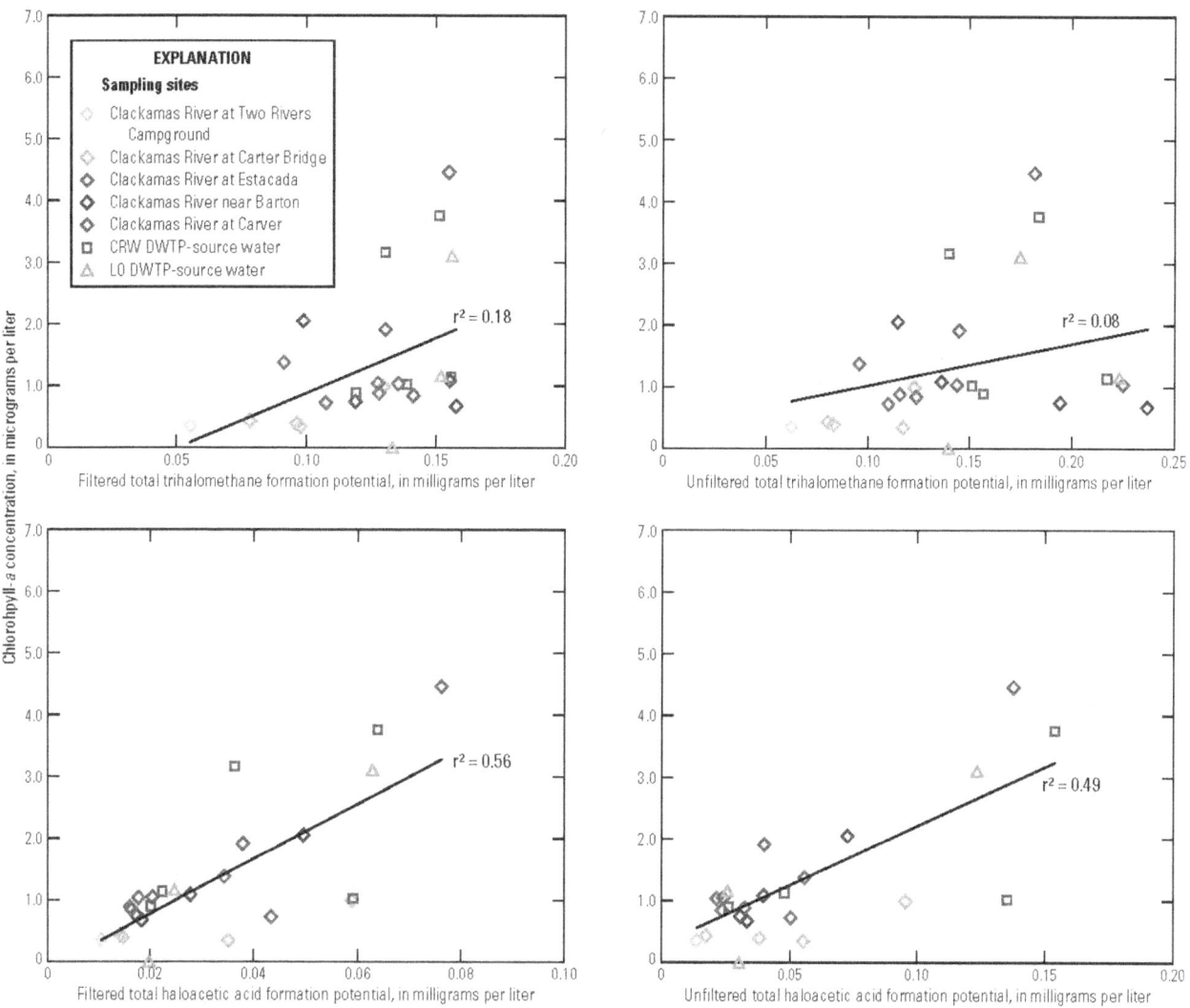

Figure 29. Relations between water-column chlorophyll-*a* and filtered and unfiltered disinfection by-product formation potentials for main-stem Clackamas River sites, Oregon, 2010–11. (Abbreviations: CRW, Clackamas River Water; LO, City of Lake Oswego.)

The relatively high STHMFP in the September 2011 sample from the release depth within North Fork Reservoir (table 9), for example, indicated the DOM present during the blue-green algae bloom was reactive and formed THMs. The amount of reactive organic matter produced by any individual bloom (and the production of algal toxins or taste and odor compounds) could depend on a number of factors, including the overall size and health of the bloom. Future studies of Timothy Lake or North Fork Reservoir could begin to characterize these processes. The concentrated "algae grab" sample with a high abundance of *Anabaena flos-aquae* (fig. 20) shows a strong signal in the lower ultraviolet "protein-like" region, which was interpreted in this system to be indicative of the more labile, freshly produced organic

matter expected in such a sample (fig. 21). This signal, represented by PARAFAC component C5, was also apparent in the full suite of reservoir samples (fig. 22). Organic matter recently contributed by these blooms is expected to contain a greater protein-like signal, particularly because *Anabaena flos-aquae* is a nitrogen-fixer and a common member of the phytoplankton assemblage in North Fork and Timothy Lake during summer.

As discussed above, algae may also contribute HAA precursors, as the positive correlations with chlorophyll-*a* suggest (fig. 29). Further, the highest HAAFP values occurred during the October 2010 storm (fig. 16), at the end of the growing season, when concentrations of chlorophyll-*a* due to algal sloughing were highest.

Phytoplankton Blooms

Blooms of blue-green algae occurred in Timothy Lake and North Fork Reservoir during this study, prompting the Oregon Health Authority to issue a human-health recreational advisory for both water bodies. Although these blooms were not as severe as in years past, elevated STHMFP values in North Fork Reservoir in September 2011 (table 9) indicate a high degree of reactivity for this type of carbon, a finding that is consistent with other studies (Jack and others, 2002; Kraus and others, 2011). Because the phytoplankton blooms in 2010 and 2011 were seemingly small, the high degree of reactivity shown by the high STHMFP suggest phytoplankton populations in the reservoirs could become important sources of THM precursors, especially if larger blooms occur in the future.

As described above, phytoplankton in the two Clackamas reservoirs could be an important source of DBP precursors for the Clackamas River. Blooms of blue-green algae and diatoms regularly occur in these reservoirs (see Carpenter, 2003), although the severity and duration vary widely, likely responding to the specific growing conditions and other factors that affect bloom development. Although sustained flow during summer in 2010 and 2011 may have limited the residence time in North Fork Reservoir, blooms were observed in the reservoir both years, but biomass was relatively low—less than 3 µg/L chlorophyll-a at the log boom. The horizontal and vertical mobility of these blooms makes it challenging to accurately characterize conditions with just one sampling. Considering longitudinal patterns, the STHMFP measurement in North Fork Reservoir was nearly two times higher compared with Carter Bridge in late summer, possibly reflecting the importance of the reservoir-derived carbon in forming THMs and possibly also contributing to taste and odor issues. The DBPFP measurements also suggested phytoplankton or their exudates produce DBP precursors. The highest STHMFP value for filtered samples, again, was from the release depth within North Fork Reservoir (table 9) and suggests the reservoir was a source of THM precursors to downstream sites.

Previous studies have examined fluxes of DBP precursors in reservoirs and lakes and found them to act as either a source or sink for DBP precursors (Stepczuk and others, 1998; Nguyen and others, 2002; Bukaveckas and others, 2007). In a recent study examining changes in DOM amount and composition in San Luis Reservoir, California, it was found that DBP precursors (particularly HAAs) increased during the summer months because of high phytoplankton activity; however, the reservoir was a sink for DBP precursors during the winter months when decomposition processes predominated (Kraus and others, 2011). Results from the Clackamas River study also suggest that information about the composition of organic matter within reservoirs, as well as its reactivity with respect to THM and HAA formation, can help explain downstream changes in source-water quality.

Benthic Algae and Periphyton Sloughing Events

Although conditions were not particularly favorable for algal growth in 2010–11, nuisance levels of benthic algae (greater than 100–150 mg/m^2; Welch and others, 1988) developed in some main-stem and tributary locations during this study (table 7). Large masses of stalked diatoms, filaments of green algae, and ears of blue-green algae (see photographs 1a-e) were common among other types of algae in the mainstem.

On the basis of water-column chlorophyll-a concentrations in the lower mainstem, which were nearly always relatively low (less than 2 µg/L), it appears on initial inspection that benthic algae were not important contributors to the water column during the study. While large algal "sloughing" events did not occur, higher-than-normal streamflow and other possible factors such as high rates of benthic invertebrate grazing or other factors may have limited or moderated water-column concentrations of chlorophyll-a during the study, as discussed above. Regardless of this finding, recognizing the unusually high streamflow, the high algal biomass observed in the river (table 7), along with the knowledge of past algal sloughing events (U.S. Geological Survey, 2012), it is suspected that sloughed benthic algae might, at times, be a significant source of DBP precursors to the lower river.

As described previously, sloughing of benthic algae is a common feature of the Clackamas River and some other Cascade Range rivers such as the North Umpqua and Rogue Rivers. Such events do not necessarily take place every year, but in some years (fig. 30), concentrations can be high for a system dominated by periphyton. Such sloughed algae impacts water clarity and fishing conditions in the Clackamas River and has clogged DWTP intake screens in the past. This demonstrates that, at times, benthic algae are a source of TOC in the river and, thus, likely a source of DBP precursors.

These events are sometimes associated with increased streamflow resulting from storms, the annual drawdown of Timothy Lake (which increases flow in the mainstem by 100 ft^3/s or more), and (or) changes in reservoir releases. For example, chlorophyll-a concentrations in the tributaries reached moderate levels (5–8 µg/L) during the October 2010 storm, which contributed to observed increases in the mainstem. The algal biomass in the mainstem more than doubled between Barton and Carver from tributary inputs— especially Deep Creek—and possibly also from benthic algae sloughed from main-stem locations downstream from Barton. Although DBP concentrations were not particularly high in finished water at the time, it raises the question about the degree to which algae may be contributing DBP precursors because chlorophyll-a levels can become elevated in the river, as this storm demonstrated.

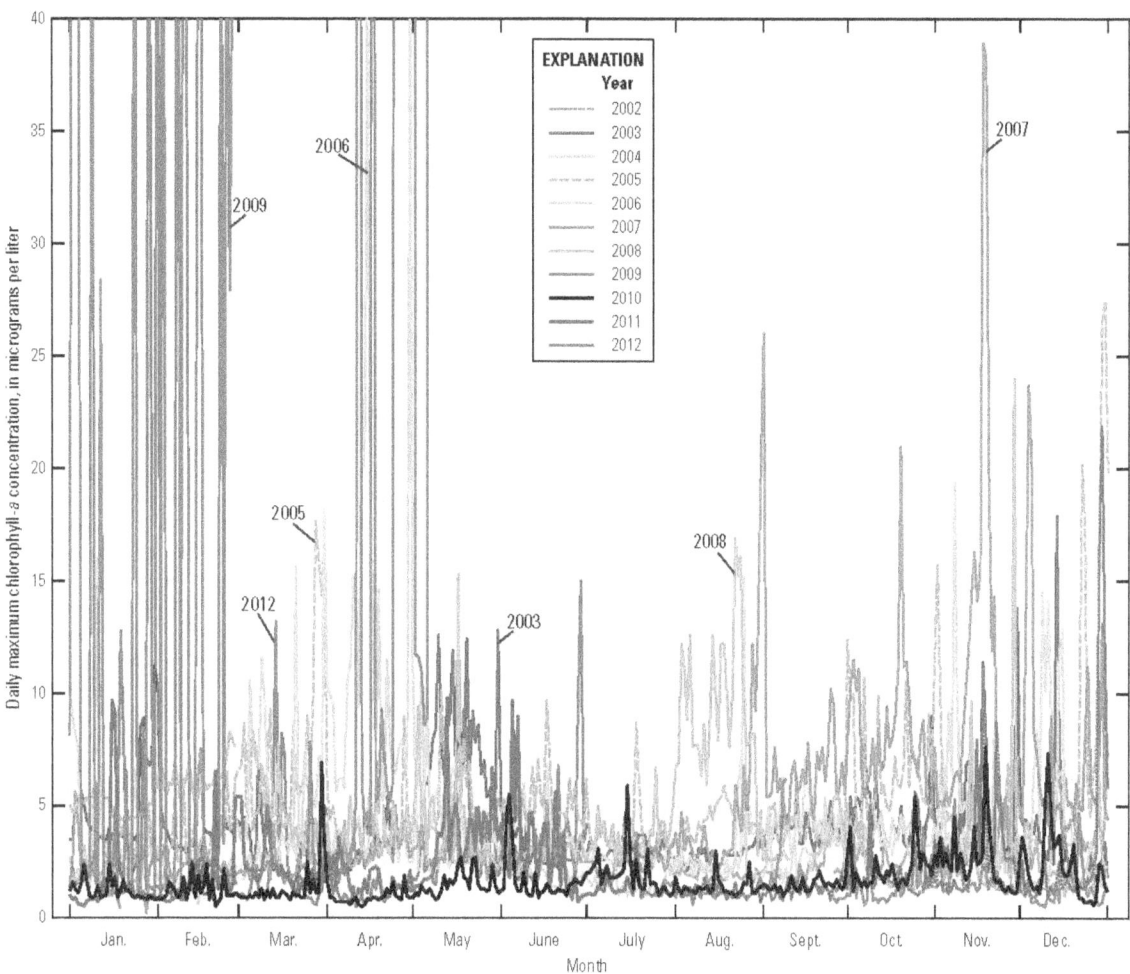

Figure 30.   Daily maximum concentrations of chlorophyll-*a* in the lower Clackamas River at Oregon City, Oregon, 2002–12, showing the higher concentrations that can result from sloughed benthic algae. (Scale is truncated at 40 micrograms per liter.)

Hydroelectric project operations, including flow ramping for power production and the annual drawdown of Timothy Lake near the end of summer starting around Labor Day, can increase the amount of algal cells, DOC, and TOC in the upper river (Carpenter, 2003). The drawdown releases phytoplankton produced through the summer growing season, and increased flows—about 10 percent higher flow in the mainstem—can scour and suspend benthic algae into the water column. Even though, as a condition of the new Federal Energy Regulatory Commission operating license, flow ramping in the upper river at the Three Lynx powerhouse, upstream from Three Lynx Creek (fig. 1), is not as great as in years past, hydroelectric project operations still cause abrupt changes in flow that can scour algae from the riverbed and increase carbon concentrations in the river.

Although not explicitly measured here, the Timothy Lake water also may contribute DBP precursors, although it did not appear to increase DBPFP substantially at Carter Bridge in this study during the drawdown (fig. 14). A comparison of the longitudinal conditions in the mainstem near the beginning of the September 2011 drawdown period to those later during the drawdown, near the peak in flow, shows the TPC and THMFP did increase in samples from Estacada downstream to the LO DWTP intake (fig. 31A). The unfiltered formation potentials were 22–81 percent higher than the filtered THMFP (not shown), suggesting a particulate source.

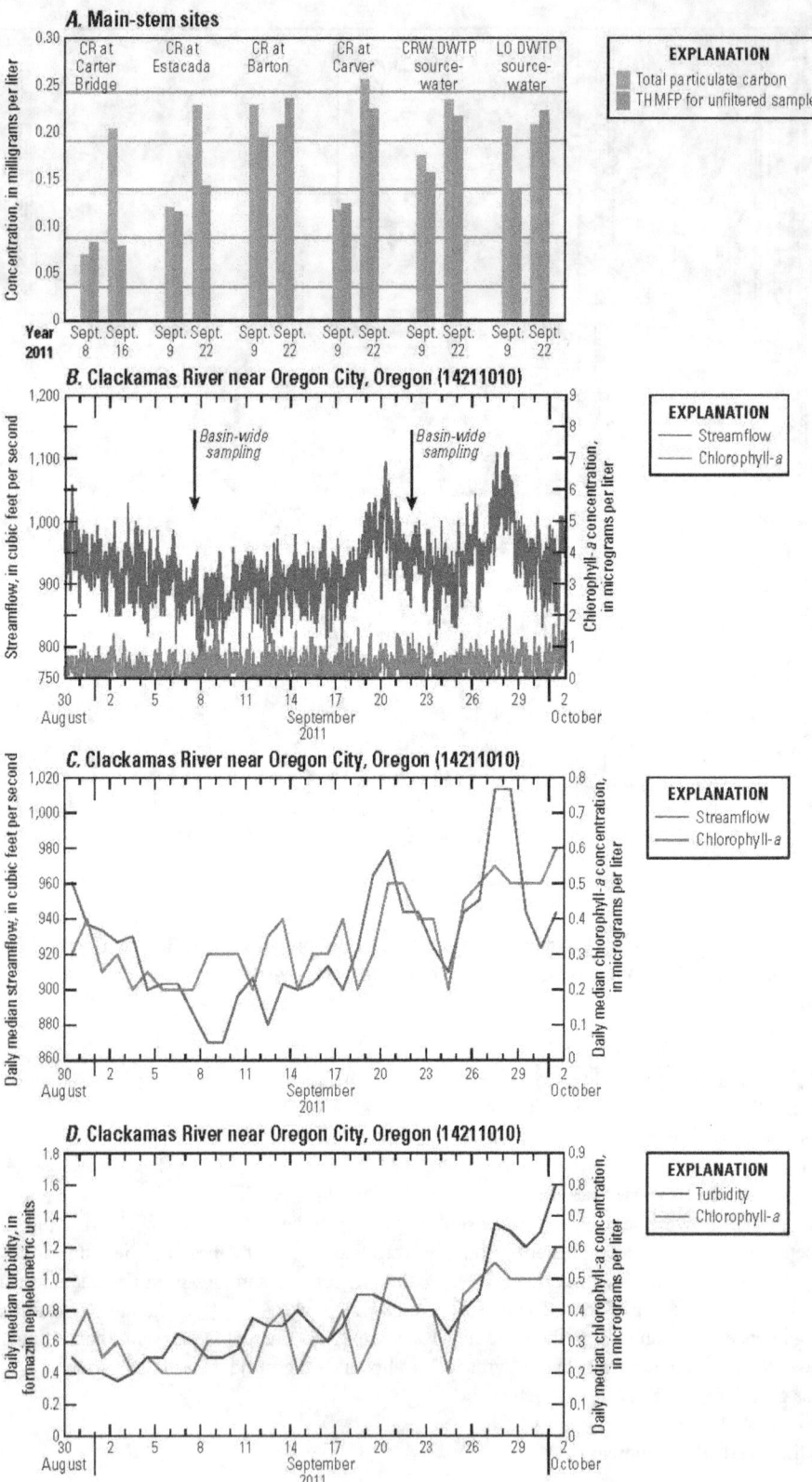

Figure 31.   September 2011 data showing (*A*) concentrations of total particulate carbon and trihalomethane formation potential for unfiltered samples in the Clackamas River at the beginning and near the peak of the Timothy Lake drawdown, (*B*) time series of streamflow and chlorophyll-*a*, (*C*) daily median values of streamflow and chlorophyll-*a*, and (*D*) daily median turbidity and chlorophyll-*a* in the lower main-stem Clackamas River at Oregon City.

Whether these September 2011 increases were due to sloughing of periphyton, or from displacement of bottom water from North Fork Reservoir (fig. 32) is not known. This "displacement" mechanism was first proposed by the USGS during the August 2003 taste and odor event at the CRW DWTP, when decreases in water temperature and increases in turbidity were noted at the Estacada water-quality monitor downstream from the North Fork Reservoir at the time of the drawdown. Taste and odor problems eventually developed in early October 2011, prompting the CRW DWTP to begin using powdered activated carbon (PAC). The cause of the event was not identified, but in August 2003, taste and odor problems were traced to geosmin, a known taste and odor compound (Graham and others, 2010), which was detected

at concentrations between 0.024 and 0.044 µg/L in North Fork Reservoir at depths of 80 and 30 ft, respectively (data furnished by CRW). Given that a bloom of blue-green algae did occur in 2011, it is certainly possible that this contributed to the observed taste and odor problem. While water displacement from North Fork Reservoir may have contributed to the 2011 taste and odor event, the cooler temperatures and higher flows in 2010–11 probably minimized this process because residence times were likely shortened and thermal temperature stratification weaker compared with 2003. Drawdown also may have enhanced the export of blue-green algae cells to the outflow in 2011 by disrupting population "layering" at depth within the reservoir, which has been noted previously for this reservoir (Carpenter, 2003).

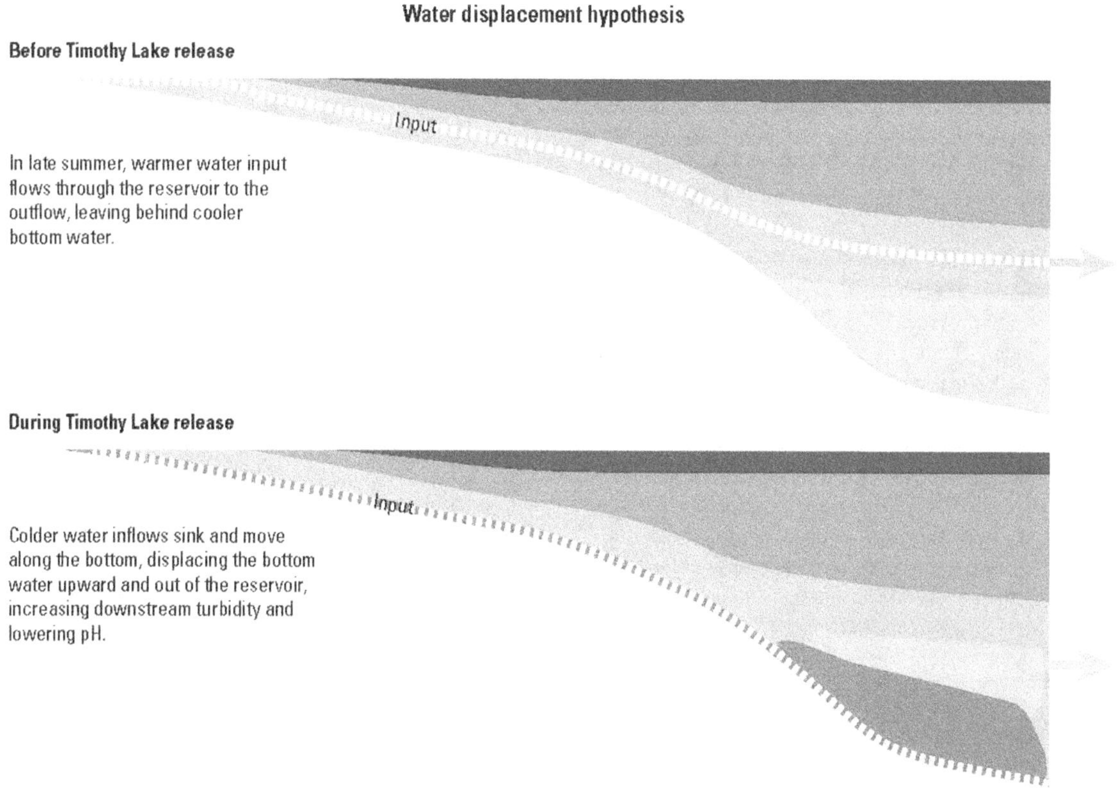

Water displacement hypothesis

Before Timothy Lake release

In late summer, warmer water input flows through the reservoir to the outflow, leaving behind cooler bottom water.

Input

During Timothy Lake release

Colder water inflows sink and move along the bottom, displacing the bottom water upward and out of the reservoir, increasing downstream turbidity and lowering pH.

Input

Figure 32. Water-displacement hypothesis, showing the proposed mechanism by which the drawdown of Timothy Lake increases transport of bottom water from North Fork Reservoir, Clackamas River basin, Oregon.

The September 2011 event also provided an opportunity to examine how water-column chlorophyll-*a* levels and turbidity respond to the increased flow in the river during the drawdown (fig. 31*B*). Even though the drawdown was one of the main differences between these dates, there was a jump in flow of 190 ft³/s in the lower river that did not cause much of a response from the chlorophyll-*a* sensor. This was not totally unexpected because it is probably the more senescent algae that would detach during the drawdown (if it were benthic algae). If blue-green algae cells were released from the reservoirs, those also would not cause a large response from the chlorophyll-*a* probe; phycocyanin probes are much better suited for detecting blue-green algae. Although the full time-series data, which has considerable within-day variation, does not show any apparent increase in chlorophyll-*a* during drawdown, the daily median values (48 daily measurements) do show a very small increase (fig. 31*C*) that may have been caused by algal sloughing. These increases were not large, however, but the small jumps in streamflow that occurred at the Oregon City streamflow gage do seem to line up, more or less, with these small increases in chlorophyll-*a*, possibly indicative of scouring. This process has been noted previously in May 2003, for example (U.S. Geological Survey, 2012). The increases in chlorophyll-*a* in 2011 were, however, quite small and probably within instrument error. While this does add to the weight of the evidence that algae are a source of DBP precursors and provides a possible mechanism that could enrich source waters, these data are not sufficient to prove or disprove the hypothesis regarding algal sloughing.

During this possible sloughing event, the THMs in finished-water samples from the CRW DWTP actually decreased, while those from the LO DWTP increased. The lower THM4 values at CRW were not, however, attributable to the use of PAC, because it was not used during the active data-collection phase of the study but was employed later in October 2011 as taste and odor problems were developing.

Wastewater

Although the Clackamas River indirectly or directly receives wastewater effluents from three wastewater treatment plants (fig. 1) and effluents from possibly thousands of septic tanks in the basin, the effluents did not produce a carbon signature that could be construed as being definitively related to wastewater. Although sampled less frequently ($n=3$), there was also no definitive wastewater signal from Deep Creek, which receives wastewater-treatment-plant discharges from the city of Sandy and town of Boring through Tickle Creek and North Fork Deep Creek, respectively.

Strategies for Managing and Reducing Disinfection By-Products

The strong terrestrial signature of the carbon in the Clackamas River suggests watershed-management strategies aimed at controlling soil erosion from forestland might reduce the input of DBP precursors to the river. Timber harvesting and associated road construction can cause erosion that commonly leads to sedimentation of streams and causes elevated turbidity. Soil erosion in the Clackamas River basin is widespread, and continuous monitoring since 2002 has shown high turbidity after storms every year (U.S. Geological Survey, 2012).

In many parts of the upper basin, unstable geology and steep slopes contribute to mass wasting and erosion, particularly in the Fish Creek and Collawash River Basins, and along portions of the upper Clackamas River where the potential for surface erosion is relatively high (Metro Regional Services, 1997). In the lower basin, tributaries become highly turbid after rain events, delivering sediment (and pesticides) to the main-stem Clackamas River (Carpenter and others, 2008); tributaries also contained high concentrations of DBP precursors. Clackamas County Water Environment Services and others are working to reduce storm-water discharge pollution, rates, and volumes in the Sieben and Rock Creek watersheds, which may reduce peak concentrations of DBP precursors in the Clackamas River (Andrew Swanson, Clackamas County Water Environment Services, written commun., 2012), although these controls were not evaluated for their efficacy in removing DBP precursors.

Certain forest-management activities such as fertilizing with urea-nitrogen, burning of post-harvest slash, and disturbing and exposing soils to direct precipitation could all contribute to leaching of DOM into the shallow groundwater system and eventual transport to the river. Although forest fertilization on Federal forestland largely ended with the Northwest Forest Plan, this practice was common until 1996 in the Clackamas River basin, where applications peaked at over 1 million pounds of nitrogen annually (Carpenter, 2003). Adams and others (2005) found that, although variable, fertilization with urea-nitrogen increased soil leaching of DOC in another Pacific Northwest Douglas-fir forest in Washington State, but it is unclear whether this has been important in the Clackamas River basin or not.

Suppression of wildfire also may increase abundance of microbes in forest soils as was found in another Douglas-fir forest in Colorado (Switzer and others, 2012). Although there have been a few small fires in the Clackamas River basin over the past 20–30 years, there have been no large fires comparable to some of the larger burns in 1930 and 1940, for example (Taylor, 1999). From 1900 to about 1940, seven large fires burned 2,000 to over 11,000 acres in the Clackamas basin (Carpenter, 2003). It is not known whether the lack of large fires has increased microbial activity or the export of carbon and DBP precursors from forested areas, but this could be an area for future study.

It is important to understand the primary sources of DBP precursors and processes that control their transport to the river. It is equally important to determine the capacity of the forest to store rather than leach organic carbon, because this may ultimately control concentrations of DBP precursors in source water and DBPs in finished drinking water, in concert with precipitation, snowpack, and other factors that affect flow and dilution rates and other influences that affect the annual growth of benthic algae and phytoplankton in this system.

Fluorescence Technology for Drinking-Water Management

Optical properties measured in the lab were highly correlated to DOC concentrations and THMFP and HAAFP in watershed samples. Similarly, in-situ FDOM sensors provided an excellent proxy for continuous DOC concentration in the Clackamas River, demonstrating great promise as a cost-effective tool to better understand DBP precursor sources, seasonality and trends, and possible management of precursor compounds. Because DOM concentration is a driving factor in the production of DPBs in finished water and a good predictor for DBPFP for all watershed samples, continuous in-situ FDOM measurements are potentially valuable as an early-warning system, and for understanding and possibly forecasting finished-water DBP concentrations. In this study, for example, FDOM was highly correlated with finished-water HAA5 concentrations, thus this relation was used to estimate continuous HAA5 concentrations over time (fig. 24).

Although both laboratory measurements of FDOM and UVA_{254} were effective proxies for DOC concentration (fig. 23 and table 11), measurement of fluorescence has several advantages over absorbance (Henderson and others, 2009; Kraus and others, 2010; Bridgeman and others, 2011). In particular, absorbance measures are more prone to metal quenching and particle interference compared with fluorescence methods, and absorbance requires filtration prior to analysis. The fact that these flat-faced and flow through fluorometers can measure fluorescence directly without filtration is a tremendous advantage. In this study, FDOM produced as high or higher correlations with finished-water DBP concentrations and laboratory DBPFPs compared with more traditional absorbance measures such as UVA_{254} (table 11). This corroborates findings from a similar study of the nearby McKenzie River (Kraus and others, 2010).

Conclusions, Implications for Drinking-Water Treatment and River Management, and Possible Future Studies

While some of the initial results of this study were used to inform recent upgrades to the water-treatment plant at the LO DWTP (Kari Duncan, City of Lake Oswego, oral commun., 2010), it is hoped that future studies build on this research and identify specific areas or activities that produce DPB precursors so that watershed-management efforts can be targeted and wisely prioritized.

The DOM precursor pool in the Clackamas River basin was strongly influenced by season, streamflow, and perhaps most notably, the effects of storms. However, not all storms were alike in terms of the amount or character of carbon exported, which reveals that factors such as time of year, antecedent flow, snowpack conditions, and patterns in rainfall and runoff affect carbon export. This study identified the primary sources of organic matter that contributed to DBP precursors in raw source-water supplies in the lower Clackamas River, which turn out to be primarily dissolved organic compounds that are terrestrial in nature. Multiple lines of evidence also supported the hypothesis that algae may, at times, be contributors to the DBP precursor pool, especially THMs.

Higher DBPFPs in some unfiltered samples compared to filtered samples suggested that, at times, a considerable amount of the total DBP precursor pool was composed of filterable particles such as detritus, soil particles, and algae. This suggests that filtration prior to chlorination could reduce finished-water DBP concentrations. Differences between watershed THM and HAA precursor sources, as well as their treatability, were evident, suggesting different actions may be necessary to manage for these types of DBP precursors.

Although terrestrial sources of carbon dominated the Clackamas River in 2010–11, some of the results obtained here suggest that algae also contributed some carbon to the river. While benthic algae reached nuisance levels in the mainstem and caused large and synchronized swings in dissolved oxygen concentrations and pH—key photosynthesis indicators—conditions did not lead to clogged intakes or substantial sloughing as has occurred years past. Such material also contributes carbon that, to some degree, contains DBP precursors. Future in-situ, high-frequency monitoring of FDOM may present opportunities not offered during this study to evaluate the degree to which algae may affect source-water quality and DBP concentrations in treated water by providing an early warning that allows for timely sampling of such conditions. Although algal blooms appear to be a regular phenomenon, weather conditions and growth–senescence processes are dynamic in the Clackamas River as algae accumulates on the riverbed or as blooms develop in the reservoirs. Sampling finished-water DBPs during periods when water-column chlorophyll-*a* is high at the Oregon City (or Estacada) monitors, or during periods of obvious sloughing of benthic algae or after particularly large blooms of blue-green algae in the reservoirs, could provide data to further evaluate this hypothesis. The occasional taste and odor event also provides opportunities to identify where in the system these compounds are coming from, and may be ideal times to screen for blue-green algae toxins (Graham and others, 2010). The high specific THMFP value from North Fork Reservoir suggests this source is worthy of future monitoring if THM concentrations continue to be a concern. Studies that examine the temporal changes in organic matter reactivity and potential production of DBP precursors, taste and odor compounds, and algal toxins over the course of a bloom could be helpful for understanding the effect that algae may have on downstream water quality. In addition to providing a better understanding of watershed processes, this knowledge might help guide future DBP-treatability approaches that remove DOC, TPC, and DBP precursors. This could yield great benefits, especially if algal-derived organic matter is less amenable to removal by coagulation methods employed at these direct-filtration plants.

High frequency, in-situ measurements of FDOM proved to be an excellent proxy for DOC concentration in the Clackamas River, suggesting further development and refinement of these sensors have the potential to provide information that can inform DWTP operations and upgrades. The in-situ FDOM measurements revealed short-term, rapid changes in DOC concentrations in the river in response to storms. For this system, the close association between source-water in-situ FDOM and finished-water HAA5 concentrations demonstrates the utility of these instruments in providing a robust proxy for DBPs continuously, in real-time. This technology represents a new tool that can be used to optimize treatment-plant operations by adjusting water-intake rates or modifying coagulant doses during critical time periods, for example, to minimize the DOC and DBP-precursor content of treated water (Kraus and others, 2010). The link between DOM fluorescence and the presence of DBP precursors for other classes of DBPs other than THMs and HAAs, including for example nitrogenous DBPs, which are of emerging concern, clearly warrants future study (Hua and others, 2007; Henderson and others, 2009; Chen and Westerhoff, 2010). Research into the use of fluorescence to monitor and predict treatability, estimate biological oxygen demand, identify and quantify wastewater inputs, and act as an early-warning system for contaminants also shows promise (Bieroza and others, 2008; 2009a, b; Hudson and others, 2008; Henderson and others, 2009; Baghoth and others, 2011; Bridgeman and others, 2011; Goldman and others, 2012).

Other studies could sample forest soils and streams to refine the current understanding of what processes lead to DOM leaching, where DBP precursors are most prevalent, and how these processes are trending over time. Information on how carbon is stored and released would help modeling and prediction of DBP precursors and be especially helpful for understanding the effects of potential changes in land management, fire, climate, precipitation patterns, and other factors. Bacterial and fungal activity within soils and decomposing wood on the forest floor likely play important roles in mediating carbon sequestration and losses over time, and more information on their status would be useful. Future studies might also examine how forest management influences forest-soil carbon dynamics and the types of organic matter (or "components" as described here) that are most prone to leaching, and which ones contribute DBP precursors. Additional measures of bromide and chloride concentrations from Austin Hot Springs in the upper basin would be helpful for detecting possible trends over time, and might provide insight into the seasonality of DBP speciation in finished drinking water.

Monitoring of in-situ DOM using high-frequency continuous FDOM sensors can provide information regarding trends in the amount and composition, and thus origin and reactivity, of organic matter present in the river and in source water. These data have not only the potential to provide real-time information that can be used to manage DWTP operations, better understand DOM treatability, and predict finished-water DBP concentrations, but also provide information about watershed hydrology and processes that affect carbon dynamics in both terrestrial and aquatic ecosystems.

Acknowledgments

Many individuals from several organizations contributed to this study. Technical and financial support came from the Water Research Foundation, the WaterRF Project Advisory Committee members, and the two water utilities that participated in the study, Clackamas River Water and the City of Lake Oswego. The authors wish to personally thank staff at CRW, especially Suzanne DeLorenzo for providing historical data from the CRW DWTP, Rob Cummings for helpful discussions regarding the water-treatment processes, and Lee Moore for his support for the project. Also at CRW, special thanks to Mike Avery and Tracy Triplett for conducting jar-test experiments and helping to service the in-situ sensors. The authors also thank and acknowledge the valuable contributions from Kari Duncan (City of Lake Oswego) for her help with project design and implementation and for providing historical data from the LO DWTP. Reliable assistance with source- and finished-water sample collection came from the DWTP operators at Clackamas River Water and City of Lake Oswego, and their efforts are greatly appreciated.

This study benefited from the six-station network of water-quality and streamflow monitors in the basin, which is supported by the Clackamas River Water Providers (CRWP),

Clackamas County Water Environment Services (CC WES), and Portland General Electric (PGE). The authors would like to thank Kimberly Swan (CRWP), Andrew Swanson (CC WES), and John Esler (PGE) for their support and help with funding this network. The authors also appreciate the contributions from Randy Kuntz, Promontory Park concessionaire, who provided daily reports on algal conditions in North Fork Reservoir and for generously providing access to a pontoon boat for sampling the reservoir.

The contributions made by staff at the Oregon and California USGS Water Science Centers are gratefully acknowledged—Stewart Rounds was instrumental in providing the software and computer programming that allowed continuous in-situ data retrieval and display in near-real-time; David Piatt, Michael Sarantou, Steven Sobieszczyk, and Tara Chestnut helped with sample collection and processing; Kathryn Crepeau conducted the DBPFP experiments; Micelis Doyle maintained the network of continuous water-quality monitors, and Rick Kittleson, Doug Cushman, Roy Wellman, Jay Spillum, Greg Lind, and Greg Olsen provided streamgage-operation and streamflow-measurement support; Matt Johnston, Amy Brooks, and Melanie North provided telecommunications and computer support; and assistance with multivariate statistical analyses using R was provided by Ian Waite. Thanks also go to Kenna Butler for her assistance with various laboratory analyses and Travis von Dessonneck for data processing.

Steven Ingebritsen (USGS, Menlo Park, CA) provided geochemical information on Austin Hot Spring, and Brian Pellerin (USGS, Sacramento, CA), Djanette Khiari (Water Research Foundation), and the Water Research Foundation Project Advisory Committee provided helpful comments on an earlier draft of the report.

References Cited

Adams, A.B., Harrison, R.B., Sletten, R.S., Strahm, B.D., Turnblom, E.C., and Jensen, C.M., 2005, Nitrogen-fertilization impacts on carbon sequestration and flux in managed coastal Douglas-fir stands of the Pacific Northwest: Forest Ecology and Management, v. 220, p. 313–325.

Aiken, G., and Cotsaris, E., 1995, Soil and hydrology—Their effect on NOM: Journal of the American Water Works Association, v. 87, no. 1, p. 36-45.

Aiken, G., Kaplan, L.A., and Weishaar, J., 2002, Assessment of relative accuracy in the determination of organic matter concentrations in aquatic systems: Journal of Environmental Monitoring, v. 4, no. 1, p. 70-74. (Also available at http://dx.doi.org/10.1039/b107322m.)

Aiken, G.R., McKnight, D.M., Thorn, K.A., and Thurman, E.M., 1992, Isolation of hydrophilic organic acids from water using nonionic macroporous resins: Organic Geochemistry, v. 18, no. 4, p. 567-573.

Andersson, C.A., and Bro, Rasmus, 2000, The *N*-way toolbox for MATLAB: Chemometrics and Intelligent Laboratory Systems, v. 52, no. 1, p. 1-4.

Baghoth, S.A., Sharma, S.K., and Amy, G.L., 2011, Tracking natural organic matter (NOM) in a drinking water treatment plant using fluorescence excitation-emission matrices and PARAFAC: Water Research, v. 45, no. 2, p. 797-809.

Beggs, K.M.H., and Summers, R.S., 2011, Character and chlorine reactivity of dissolved organic matter from a mountain pine beetle impacted watershed: Environmental Science & Technology, v. 45, p. 5717–5724.

Beggs, K.M.H., Summers, R.S., and McKnight, D.M., 2009, Characterizing chlorine oxidation of dissolved organic matter and disinfection by-product formation with fluorescence spectroscopy and parallel factor analysis: Journal of Geophysical Research-Biogeosciences, v. 114, G04001, 10 p.

Bergamaschi, B.A., Kalve, Erica, Guenther, Larry, Mendez, G.O., and Belitz, Kenneth, 2005, An assessment of optical properties of dissolved organic material as quantitative source indicators in the Santa Ana River Basin, Southern California: U.S. Geological Survey Scientific Investigations Report 2005-5152, 38 p.

Bergamaschi, B.A., Krabbenhoft, D.P., Aiken, G.R., Patino, Eduardo, Rumbold, D.G., and Orem, W.H., 2012, Tidally driven export of dissolved organic carbon, total mercury, and methylmercury from a mangrove-dominated estuary: Environmental Science & Technology, v. 46, no. 3, p. 1371-1378. (Also available at http://dx.doi.org/10.1021/es2029137.)

Bieroza, M., Baker, A., and Bridgeman, J., 2009a, Relating freshwater organic matter fluorescence to organic carbon removal efficiency in drinking water treatment: Science of the Total Environment, v. 407, no. 5, p. 1765-1774.

Bieroza, M., Baker, A., and Bridgeman, J., 2009b, Exploratory analysis of excitation-emission matrix fluorescence spectra with self-organizing maps as a basis for determination of organic matter removal efficiency at water treatment works: Journal of Geophysical Research-Biogeosciences, v. 114. (Also available at http://dx.doi.org/10.1029/2009JG000940.)

Blough, N.V., and Del Vecchio, R., 2002, Chromophoric DOM in the coastal environment, *in* Hansell, D.A., and Carlson, C.A., eds., Biogeochemistry of Marine Dissolved Organic Matter: Academic Press, San Diego, Calif., p. 509–546.

Bridgeman, J., Bieroza, M., and Baker, A., 2011, The application of fluorescence spectroscopy to organic matter characterisation in drinking water treatment: Reviews in Environmental Science and Bio-Technology, v. 10, no. 3, p. 277-290.

Bro, Rasmus, 1997, PARAFAC—Tutorial and applications: Chemometrics and Intelligent Laboratory Systems, v. 38, no. 2, p. 149-171. (Also available at http://dx.doi.org/10.1016/S0169-7439(97)00032-4.)

Bukaveckas, P.A., McGaha, Dale, Shostell, J.M., Schultz, Richard, and Jack, J.D., 2007, Internal and external sources of THM precursors in a midwestern reservoir: Journal of the American Water Works Association, v. 99, no. 5, p. 127-136.

Carpenter, K.D., 2003, Water-quality and algal conditions in the Clackamas River basin, Oregon, and their relations to land and water management: U.S. Geological Survey Water-Resources Investigations Report 02-4189, 114 p.

Carpenter, K.D., and McGhee, Gordon, 2009, Organic compounds in Clackamas River water used for public supply near Portland, Oregon, 2003–05: U.S. Geological Survey Fact Sheet 2009–3030, 6 p.

Carpenter, K.D., Sobieszczyk, Steven, Arnsberg, A.J., and Rinella, F.A., 2008, Pesticide occurrence and distribution in the lower Clackamas River basin, Oregon, 2000–2005: U.S. Geological Survey Scientific Investigations Report 2008–5027, 98 p.

Chen, B., and Westerhoff, P., 2010, Predicting disinfection by-product formation potential in water: Water Research, v. 44, no. 13, p. 3755-3762.

Chen C., Zhang, X.J., Zhu, L.X., Liu, J., He, W.J., Han, H.D., 2008, Disinfection by-products and their precursors in a water treatment plant in North China—Seasonal changes and fraction analysis: Science of the Total Environment, v. 397, p. 140–147.

Clarke, K.R., and Gorley, R.N., 2006, PRIMER v. 6, User Manual: Plymouth, U.K., Primer-E-Ltd., 190 p.

Coble, P.G., 2007, Marine optical biogeochemistry—The chemistry of ocean color: Chemical Reviews, v. 107, no. 2, p. 402-418. (Also available at http://dx.doi.org/10.1021/cr050350+.)

Cooke, G.D., and Kennedy, R.H., 2001, Managing drinking water supplies: Journal of Lake and Reservoir Management, v. 17, no. 3, p. 157-174.

Cory, R.M., and McKnight, D.M., 2005, Fluorescence spectroscopy reveals ubiquitous presence of oxidized and reduced quinones in dissolved organic matter: Environmental Science and Technology, v. 39, p. 8142–8149.

Cory, R.M., Miller, M.P., McKnight, D.M., Guerard, J.J., and Miller, P.L., 2010, Effect of instrument-specific response on the analysis of fulvic acid fluorescence spectra: Limnology and Oceanography-Methods, v. 8, p. 67-78.

Crepeau, K.L., Fram, M.S., and Bush, Noël, 2004, Method of analysis at the U.S. Geological Survey California Sacramento Laboratory—Determination of trihalomethane formation potential, method validation, and quality-control practices: U.S. Geological Survey Scientific Investigations Report 2004-5003, 21 p.

Croué, J.P., DeBroux, J.F., Amy, G.L., Aiken, G.R., and Leenheer, J.A., 1999, Natural organic matter—Structural characteristics and reactive properties, in Singer, P.C., ed., Formation and control of disinfection by-products in drinking water: Denver, Colo., American Water Works Association, p. 65-93.

Croué, J.P., Violleau, D., and Labouyrie, L., 2000, Disinfection by-product formation potentials of hydrophobic and hydrophilic natural organic matter fractions—A comparison between a low- and high-humic water, in Barrett, S.E., Krasner, S.W., and Amy, G.L., eds., Natural organic matter and disinfection by-products—Characterization and control in drinking water: Washington, D.C., American Chemical Society, American Chemical Society Symposium Series 761, p. 139-153.

DeRoo, T.G., Smith, Doug, and Anderson, Doug, 1998, Factors affecting landslide incidence after large storm events during the winter of 1995-1996 in the upper Clackamas River drainages, Oregon Cascades, in Burns, S., ed., Environmental, groundwater and engineering geology—Applications from Oregon: Belmont, Calif., Star Publishing Company, p. 379-390.

Downing, B.D., Bergamaschi, B.A., Evans, D.G., and Boss, Emmanuel, 2008, Assessing contribution of DOC from sediments to a drinking-water reservoir using optical profiling: Lake and Reservoir Management, v. 24, no. 4, p. 381-391.

Downing, B.D., Boss, Emmanuel, Bergamaschi, B.A., Fleck, J.A., Lionberger, M.A., Ganju, N.K., Schoellhamer, D.H., and Fujii, Roger, 2009, Quantifying fluxes and characterizing compositional changes of dissolved organic matter in aquatic systems in situ using combined acoustic and optical measurements: Limnology and Oceanography—Methods, v. 7, p. 119-131. (Also available at http://dx.doi.org/10.4319/lom.2009.7.119.)

Downing, B.D., Pellerin, B.A., Saraceno, J.F., Bergamaschi, B.A., and Kraus, T.E.C., 2012, Seeing the light—The effects of temperature, inner filtering and particles, on in situ measurements of DOM fluorescence in rivers and streams: Limnology and Oceanography—Methods 10, p. 767–775.

Edwards, T.K., and Glysson, D.G., 1999, Field methods for measurement of fluvial sediment: U.S. Geological Survey Techniques of Water-Resources Investigations, book 3, chap. C2, 80 p.

Edzwald, J.K., Becker, W.C., and Wattier, K.L., 1985, Surrogate parameters for monitoring organic matter and THM precursors: Journal of American Water Works Association, v. 77, no. 4, p. 122-132.

Fellman, J.B., Hood, E., and Spencer, R.G.M., 2010, Fluorescence spectroscopy opens new windows into dissolved organic matter dynamics in freshwater ecosystems—A review: Limnology and Oceanography, v. 55, no. 6, p. 2452–2462. (Also available at http://dx.doi. org/10.4319/lo.2010.55.6.2452.)

Goldman, J.H., Rounds, S.A., and Needoba, J.A., 2012, Applications of fluorescence spectroscopy for predicting percent wastewater in an urban stream: Environmental Science and Technology, v. 46, no. 8, p. 4374–4387.

Graham, J.L, Loftin, K.A., Meyer, M.T., and Ziegler, A.C., 2010, Cyanotoxin mixtures and taste-and-odor compounds in cyanobacterial blooms from the midwestern United States: Environmental Science and Technology, v. 44, no. 19, p. 7361–7368.

Graham, N.J.D., Wardlaw, V.E., Perry, R., and Jiang, J.Q., 1998, The significance of algae as trihalomethane precursors: Water Science and Technology, v. 37, no. 2, p. 83-89.

Helms, J.R., Stubbins, Aron, Ritchie, J.D., Minor, E.C., Kieber, D.J., and Mopper, Kenneth, 2008, Absorption spectral slopes and slope ratios as indicators of molecular weight, source, and photobleaching of chromophoric dissolved organic matter: Limnology and Oceanography, v. 53, no. 3, p. 955-969.

Henderson, R.K., Baker, A., Murphy, K.R., Hambly, A., Stuetz, R.M., and Kahn, S.J., 2009, Fluorescence as a potential monitoring tool for recycled water systems—A review: Water Research, v. 43, no. 4, p. 863-881.

Hernes, P.J., Bergamaschi, B.A., Eckard, R.S., and Spencer, R.G.M., 2009, Fluorescence-based proxies for lignin in freshwater dissolved organic matter: Journal of Geophysical Research-Biogeosciences, v. 114, G00F03, 10 p.

Hong, H.C., Mazumder, Asit, Wong, M.H., and Liang, Yan, 2008, Yield of trihalomethanes and haloacetic acids upon chlorinating algal cells, and its prediction via algal cellular biochemical composition: Water Research, v. 42, no. 20, p. 4941-4948.

Hua, Bin, Veum, Kristen, Koirala, Amod, Jones, John, Clevenger, Thomas, and Deng, Baolin, 2007, Fluorescence fingerprints to monitor total trihalomethanes and N-nitrosodimethylamine formation potentials in water: Environmental Chemistry Letters, v. 5, no. 2, p. 73-77.

Hua, Bin, Veum, Kristen, Yang, John, Jones, John, and Deng, Baolin., 2010, Parallel factor analysis of fluorescence EEM spectra to identify THM precursors in lake waters: Environmental Monitoring and Assessment, v. 161, no. 1-4, p. 71-81.

Huang, J., Graham, N., Templeton, M.R., Zhang, Y., Collins, C., and Nieuwenhuijsen, M., 2009, A comparison of the role of two blue-green algae in THM and HAA formation: Water Research, v. 43, no. 12, p. 3009-3018.

Hudson, N., Baker, A., and Reynolds, D., 2007, Fluorescence analysis of dissolved organic matter in natural, waste and polluted waters—A review: River Research and Applications, v. 23, p. 631-649. (Also available at http:// dx.doi.org/10.1002/rra.1005.)

Hudson, N., Baker, A., Ward, D., Reynlds, D.M., Brunsdon, C., Carliell-Marquet, C., and Browning, S., 2008, Can fluorescence spectrometry be used as a surrogate for the Biochemical Oxygen Demand (BOD) test in water quality assessment? An example from South West England: Science of the Total Environment, v. 391, no. 1, p. 149-158.

Jack, Jeffrey, Sellers, Tim, and Bukaveckas, P.A., 2002, Algal production and trihalomethane formation potential—An experimental assessment and inter-river comparison: Canadian Journal of Fisheries and Aquatic Sciences, v. 59, no. 9, p. 1482-1491. (Also available at http://dx.doi. org/10.1139/F02-121.)

Kitis, M., Karanfil, T., Kilduff, J.E., and Wigton, A., 2001, The reactivity of natural organic matter to disinfection byproducts formation and its relation to specific ultraviolet absorbance: Water Science and Technology, v. 43, no. 2, p. 9-16.

Korshin, G.V., Li, C.W., and Benjamin, M.M., 1997, Monitoring the properties of natural organic matter through UV spectroscopy—A consistent theory: Water Research, v. 31, no. 7, p. 1787-1795.

Krasner, S.W., Weinberg, H.S., Richardson, S.D., Pastor, S.J., Chinn, R., Sclimenti, M.J., Onstad, G.D., and Thurston, A.D., Jr., 2006, Occurrence of a new generation of disinfection byproducts: Environmental Science and Technology, v. 40, no. 23, p. 7175-7185.

Kraus, T.E.C., Anderson, C.A., Morgenstern, Karl, Downing, B.D., Pellerin, B.A., and Bergamaschi, B.A., 2010, Determining sources of dissolved organic carbon and disinfection byproduct precursors to the McKenzie River, Oregon: Journal of Environmental Quality, v. 39, no. 6, p. 2100-2112. (Also available at http://dx.doi.org/10.2134/jeq2010.0030.)

Kraus, T.E.C., Bergamaschi, B.A., Hernes, P.J., Doctor, D., Kendall, C., Downing, B.D., and Losee, R.F., 2011, How reservoirs alter drinking water quality—Organic matter sources, sinks, and transformations: Lake and Reservoir Management, v. 27, no. 3, p. 205-219.

Kraus, T.E.C., Bergamaschi, B.A., Hernes, P.J., Spencer, R.G.M., Stepanauskas, R., Kendall, C., Losee, R.F., and Fujii, R., 2008, Assessing the contribution of wetlands and subsided islands to dissolved organic matter and disinfection byproduct precursors in the Sacramento-San Joaquin River Delta—A geochemical approach: Organic Geochemistry, v. 39, no. 9, p. 1302-1318.

Lakowicz, J.R., 2006, Principles of fluorescence spectroscopy (3rd ed.): Springer Science and Business Media, p. 954

Levesque, Steven, Rodriguez, M.J., Serodes, Jean, Beaulieu, Christine, and Proulx, François, 2006, Effects of indoor drinking water handling on trihalomethanes and haloacetic acids: Water Research, v. 40, no. 15, p. 2921–2930.

Liang, L., and Singer, P.C., 2003, Factors influencing the formation and relative distribution of haloacetic acids and trihalomethanes in drinking water: Environmental Science and Technology, v. 37, no. 13, p. 2920-2928.

Marhaba, T.F., Borgaonkar, A.D., and Punburananon, Krit, 2009, Principal component regression model applied to dimensionally reduced spectral fluorescent signature for the determination of organic character and THM formation potential of source water: Journal of Hazardous Materials, v. 169, no. 1-3, p. 998-1004.

Matilainen, Anu, Gjessing, E.T., Lahtinen, Tanja, Hed, Leif, Bhatnagar, Amit, and Sillanpää, Mika, 2011, An overview of the methods used in the characterisation of natural organic matter (NOM) in relation to drinking water treatment: Chemosphere, v. 83, no. 11, p. 1431-1442.

McCulloch, A., 2002, Trichloroacetic acid in the environment—A Review: Chemosphere, v. 47, no. 7, p. 667–686.

McKnight, D.M., Boyer, E.W., Westerhoff, P.K., Doran, P.T., Kulbe, Thomas, and Andersen, D.T., 2001, Spectrofluorometric characterization of dissolved organic matter for indication of precursor organic material and aromaticity: Limnology and Oceanography, v. 46, no. 1, p. 38-48.

Metro Regional Services, 1997, Clackamas River Watershed atlas: Portland, Oregon, Metro Regional Services, 41 p.

Moulton, S.R., II, Kennen, J.G., Goldstein, R.M., and Hambrook, J.A., 2002, Revised protocols for sampling algal, invertebrate, and fish communities as part of the national water quality assessment program: U.S. Geological Survey Open-File Report 02–150, 72 p.

Murphy, K.R., Stedmon, C.A., Waite, T.D., and Ruiz, G.M., 2008, Distinguishing between terrestrial and autochthonous organic matter sources in marine environments using fluorescence spectroscopy: Marine Chemistry, v. 108, no. 1-2, p. 40-58. (Also available at http://dx.doi.org/10.1016/j.marchem.2007.10.003.)

Nakajima, F., Hanabusa, M., and Furumai, H., 2002, Excitation–emission fluorescence spectra and trihalomethane formation potential in the Tama River, Japan: Water Science and Technology—Water Supply, v. 2, no. 5-6, p. 481–486.

Nguyen, M.L., Baker, L.A., and Westerhoff, P., 2002, DOC and DBP precursors in western US watersheds and reservoirs: Journal of the American Water Works Association, v. 94, no. 5, p. 98-112.

Nguyen, M.L., Westerhoff, Paul, Baker, Lawrence, Hu, Qiang, Esparza-Soto, Mario, and Sommerfeld, Milton, 2005, Characteristics and reactivity of algae-produced dissolved organic carbon: Journal of Environmental Engineering, v. 131, no. 11, p. 1574-1582. (Also available at http://dx.doi.org/10.1061/(ASCE)0733-9372(2005)131:11(1574).)

Obernosterer I., and Benner, R., 2004, Competition between biological and photochemical processes in the mineralization of dissolved organic carbon: Limnology and Oceanography, v. 49, p.117–124.

Ohno, Tsutomu, 2002, Fluorescence inner-filtering correction for determining the humification index of dissolved organic matter: Environmental Science and Technology, v. 36, p. 742-746.

Oregon Health Authority, 2012, Drinking water data online: Portland, Oregon, Drinking Water Services, accessed April 2, 2012, at http://170.104.63.9/.

Pellerin, B.A., Saraceno, J.F., Shanley, J.B., Sebestyen, S.D., Aiken, G.R., Wollheim, W.M., and Bergamaschi, B.A., 2012, Taking the pulse of snowmelt—in situ sensors reveal seasonal, event and diurnal patterns of nitrate and dissolved organic matter variability in an upland forest stream: Biogeochemistry, v. 108, no. 1-3, p. 183-198. (Also available at http://dx.doi.org/10.1007/s10533-011-9589-8.)

Piper, A.M., 1942, Ground-water resources of the Willamette Valley, Oregon: U.S. Geological Survey Water-Supply Paper 890, 194 p.

Rathbun, R.E., 1996, Trihalomethane and nonpurgeable total organic-halide formation potentials of the Mississippi River: Archives of Environmental Contamination and Toxicology, v. 30, no. 2, p. 156-162.

Reckhow, D.A., Rees, P.L.S., and Bryan, D., 2004, Watershed sources of disinfection byproduct precursors: Water Science and Technology—Water Supply, v. 4, no. 4, p. 61-69.

Richardson, S.D., Plewa, M.J., Wagner, E.D., Schoeny, Rita, and DeMarini, D.M., 2007, Occurrence, genotoxicity, and carcinogenicity of regulated and emerging disinfection by-products in drinking water—A review and roadmap for research: Mutation Research, v. 636, no. 1-3, p. 178-242.

Sadiq, R., and Rodriguez, M.J., 2004, Disinfection by-products (DBPs) in drinking water and predictive models for their occurrence: a review: Science of the Total Environment, v. 321, no. 1-3, p. 21-46.

Sakamoto, C.M., Friederich, G.E., and Codispoti, L.A., 1990, MBARI procedures for automated nutrient analyses using a modified Alpkem Series 300 Rapid Flow Analyzer: Moss Landing, Calif., Monterey Bay Aquarium Research Institute, MBARI Technical Report 90-2.

Saraceno, J.F., Pellerin, B.A., Downing, B.D., Boss, Emmanuel, Bachand, P.A.M., and Bergamaschi, B.A., 2009, High-frequency in situ optical measurements during a storm event—Assessing relationships between dissolved organic matter, sediment concentrations, and hydrologic processes: Journal of Geophysical Research-Biogeosciences, v. 114, G00F09, 11 p. (Also available at http://dx.doi.org/10.1029/2009JG000989.)

SAS Institute, Inc., 2003, The analyst application (2d ed.): Cary, N.C., SAS Institute Inc., 480 p.

Sharp, E.L., Parsons, S.A., and Jefferson, Bruce, 2006, The impact of seasonal variations in DOC arising from a moorland peat catchment on coagulation with iron and aluminum salts: Environmental Pollution, v. 140, no. 3, p. 436-443. (Also available at http://dx.doi.org/10.1016/j.envpol.2005.08.001.)

Shin, J.Y., Spinette, R.F., and O'Melia, C.R., 2008, Stoichiometry of coagulation revisited: Environmental Science & Technology, v. 42, no. 7, p. 2582-2589.

Spencer, R.G.M., Pellerin, B.A., Bergamaschi, B.A., Downing, B.D., Kraus, T.E.C., Smart, D.R., Dahgren, R.A., and Hernes, P.J., 2007, Diurnal variability in riverine dissolved organic matter composition determined by in situ optical measurement in the San Joaquin River (California, USA): Hydrological Processes, v. 21, no. 23, p. 3181-3189. (Also available at http://dx.doi.org/10.1002/hyp.6887.)

Spencer, R.G.M., Stubbins, Aron, Hernes, P.J., Baker, Andy, Mopper, Kenneth, Aufdenkampe, A.K., Dyda, R.Y., Mwamba, V.L., Mangangu, A.M., Wabakanghanzi, J.N., and Six, Johan, 2009, Photochemical degradation of dissolved organic matter and dissolved lignin phenols from the Congo River: Journal of Geophysical Research-Biogeosciences, v. 114, 12 p. (Also available at http://dx.doi.org/10.1029/2009JG000968.)

Stedmon, C.A., and Bro, Rasmus, 2008, Characterizing dissolved organic matter fluorescence with parallel factor analysis—A tutorial: Limnology and Oceanography—Methods, v. 6, p. 572-579. (Also available at http://dx.doi.org/10.4319/lom.2008.6.572.)

Stedmon, C.A., and Markager, Stiig, 2005, Resolving the variability of dissolved organic matter fluorescence in a temperate estuary and its catchment using PARAFAC analysis: Limnology and Oceanography, v. 50, no. 2, p. 686-697. (Also available at http://dx.doi.org/10.4319/lo.2005.50.2.0686.)

Stedmon, C.A., Markager, S., and Bro, R., 2003, Tracing dissolved organic matter in aquatic environments using a new approach to fluorescence spectroscopy: Marine Chemistry, v. 82, no. 3-4, p. 239-254.

Stepczuk, C., Martin, A.B., Longabucco, P., Bloomfield, J.A., and Effler, S.W., 1998, Allochthonous contributions of THM precursors in a eutrophic reservoir: Journal of Lake and Reservoir Management, v. 14, no. 2-3, p. 344-355.

Summers, R.S., Hooper, S.M., Shukairy, H.M., Solarik, Gabriele, and Owen, Douglas, 1996, Assessing the DBP yield—Uniform formation conditions: Journal of the American Water Works Association, v. 88, no. 6, p. 80-93.

Switzer, J.M., Hope, G.D., Grayston, S.J., and Prescott, C.E., 2012, Changes in soil chemical and biological properties after thinning and prescribed fire for ecosystem restoration in a Rocky Mountain Douglas-fir forest: Forest Ecology and Management, v. 275, p. 1–13.

Taylor, Barbara, 1999, Salmon and steelhead runs and related events of the Clackamas River basin—A historical perspective: Portland, Oregon, Portland General Electric IPS2-17188, 59 p.

Turner, D.P., Ritts, W.D., Yang, Z., Kennedy, R.E., Cohen, W.B., Duane, M.V., Thornton, P.E., and Law, B.E., 2011, Decadal trends in net ecosystem production and net ecosystem carbon balance for a regional socioecological system: Forest Ecology Management, v. 262, no. 7, p. 1318-1325.

Twardowski, M.S., Boss, Emmanuel, Sullivan, J.M., and Donaghay, P.L., 2004, Modeling the spectral shape of absorption by chromophoric dissolved organic matter: Marine Chemistry, v. 89, no. 1-4, p. 69-88. (Also available at http://dx.doi.org/10.1016/j.marchem.2004.02.008.)

U.S. Environmental Protection Agency, 2005, Drinking water criteria document for brominated trihalomethanes: Washington, D.C., Office of Water Report EPA-822-R-05-011, accessed March 30, 2012, at http://water.epa.gov/action/advisories/drinking/upload/2006_05_04_criteria_drinking_brthm-summary-200605.pdf.

U.S. Environmental Protection Agency, 2006, National primary drinking water regulations—Stage 2 disinfectants and disinfection byproducts rule: Federal Register v. 71, no. 18, accessed April 24, 2012, at https://www.federalregister.gov/articles/2006/01/04/06-3/national-primary-drinking-water-regulations-stage-2-disinfectants-and-disinfection-byproducts-rule.

U.S. Environmental Protection Agency, 2009, National primary drinking water regulations—Minor correction to stage 2 disinfectants and disinfection byproducts rule and changes in references to analytical methods: Federal Register, June 29, 2009, v. 74, no. 123, accessed April 24, 2012, at http://www.epa.gov/fedrgstr/EPA-WATER/2009/June/Day-29/w14598.htm.

U.S. Geological Survey, 2011, USGS data grapher and data tabler: U.S. Geological Survey Web site, accessed December 12, 2012, at http://or.water.usgs.gov/grapher/.

U.S. Geological Survey, 2012, Clackamas River water quality monitors: Portland, Oregon, Oregon Water Science Center, accessed April 24, 2012 at http://or.water.usgs.gov/clackamas/monitors/.

Weishaar, J.L., Aiken, G.R., Bergamaschi, B.A., Fram, M.S., Fujii, Roger, and Mopper, Kenneth, 2003, Evaluation of specific ultraviolet absorbance as an indicator of the chemical composition and reactivity of dissolved organic carbon: Environmental Science & Technology, v. 37, no. 20, p. 4702-4708.

Welch, E.B., Jacoby, J.M., Horner, R.R., and Seeley, M.R., 1988, Nuisance biomass levels of periphytic algae in streams: Hydrobiologia, v. 157, p. 161-168.

Yamashita, Y., and Tanoue, E., 2003, Chemical characterization of protein-like fluorophores in DOM in relation to aromatic amino acids: Marine Chemistry, v. 82, no. 3, p. 255-271. (Also available at http://dx.doi.org/10.1016/S0304-4203(03)00073-2.)

Yamashita, Youhei, Jaffe, Rudolf, Maie, Nagamitsu, and Tanoue, Eiichiro, 2008, Assessing the dynamics of dissolved organic matter (DOM) in coastal environments by excitation emission matrix fluorescence and parallel factor analysis (EEM-PARAFAC): Limnology and Oceanography, v. 53, no. 5, p. 1900-1908. (Also available at http://dx.doi.org/10.4319/lo.2008.53.5.1900.)

Zepp, R.G., Sheldon, W.M., and Moran, M.A., 2004, Dissolved organic fluorophores in southeastern US coastal waters—Correction method for eliminating Rayleigh and Raman scattering peaks in excitation–emission matrices: Marine Chemistry, v. 89, no. 1–4, p. 15-36. (Also available at http://dx.doi.org/10.1016/j.marchem.2004.02.006.)

Zsolnay, A., Baigar, E., Jimenez, M., Steinweg, B., and Saccomandi, F., 1999, Differentiating with fluorescence spectroscopy the sources of dissolved organic matter in soils subjected to drying: Chemisphere, v. 38, p. 45–50.

Appendixes and Data Quality Assurance

Evaluation of Quality-Assurance Data

Quality-assurance (QA) water samples consisted of field and laboratory blanks, field and laboratory (churn split) replicates, matrix spike samples, and standard reference samples. Appendix A lists the various types of QA samples for each type of data; appendix B provides the QA results for dissolved organic carbon (DOC) and optics; appendix C shows QA data for disinfection by-products (DBPs) in finished water; appendix D shows QA data for DBP formation potential (DBPFP) measurements; appendix E shows QA results for nutrients.

Organic carbon and optical properties—QA samples for carbon and optical properties included four laboratory replicates (churn split) and two field replicates (appendix B). While the average relative percent differences (RPDs) for replicate samples were within 5–6 percent for total fluorescence, DOC, ultraviolet absorbance (UVA), fluorescence index (FI), and four of five carbon component loadings, higher average RPDs were found for component C5 (11 percent), spectral slopes (8–11 percent), humic index (HIX) and peaks B, T, and N (20–36 percent). The one finished-water replicate had some relatively high RPDs (appendix B), but in many cases, these represented small actual differences in concentrations that were at or near the detection limit.

More variability in C5 values produced higher standard deviations compared with other components, possibly because the protein-like organic matter is highly variable amongst sites and seasonal trends (Yamashita and Tanoue, 2003). Reproducibility and precision of measurements of protein-like peaks in natural waters are reduced at shorter wavelength, because of interference in background fluorescence (Yamashita and Tanoue, 2003). Instability of the protein peaks can be due to changes in the complex structures of the protein. The fluorescing component of proteins found in organic matter resides in the residuals of the protein folds. Tryptophan and tyrosine display high anisotropies that are usually sensitive to protein conformation and the extent of motion during the excited state. This leads to highly variable natural lifetimes of proteins and makes it difficult to have accuracy in the reproducibility of the fluorescence of these particular fluorophores.

DBP—QA samples for DBPs included "in-house" samples prepared by the laboratory and one blank, three replicates, and one standard reference sample submitted blindly during the project. Each set of 10 samples analyzed by the laboratory included 1 blank sample, pre- and post-CCV (continuing calibration verification) samples, and 1 matrix-spiked sample (regular sample spiked with 0.01 mg/L).

The 1 blank sample submitted blindly contained no detections, and 12 in-house laboratory blanks for chloroform and bromodichloromethane also resulted in no detections. The standard reference sample for DBPs indicated a high percent relative difference for a few compounds, notably bromoform and chloroform (appendix C). The percent recovery for chloroform was just 87 percent, and although this may represent a low bias, this represented only a small difference between expected and reported concentrations of 0.01 mg/L (appendix C). In-house matrix spike samples (0.01 mg/L) for these two compounds resulted in average percent recoveries of 98±13 percent and 104±9 percent, respectively, whereas recovery ranges were 74–125 percent and 78–117 percent. The RPDs among replicate spike samples were about 3 percent for both compounds (Adriana Gonzalez-Gray, Alexin Laboratories, written commun., 2012). Field replicate values showed more variation in trihalomethanes (THMs) (notably chloroform) compared with haloacetic acids (HAAs), although absolute differences between replicates was again small, 0.001–0.004 mg/L (appendix C).

DBP Formation Potentials—One blank sample contained a low-level concentration of chloroform at the detection limit of 0.002 mg/L (appendix B). Given the much higher concentrations present in environmental samples, this low-level detection does not affect the results or interpretations. Field DBPFP replicates showed variations of 5–10 percent for THMs and 5–20 percent for HAAs; the highest relative differences were for trichloroacetic acid (TCAA [appendix D]). Samples not meeting the upper acceptable range of 30 percent generally had low DPB concentrations (less than 0.010 mg/L), thus the absolute differences were small. Differences between field duplicates were largest for chloroform differences between field duplicates were largest for chloroform, which is more volatile compared with the other DBPs, and thus most prone to loss due to the occasional air bubble. The occurrence of bubbles in some DBPFP samples may have resulted in a low bias for the highly volatile compounds such as chloroform, but as the standard reference results presented below show, this does not appear to be a widespread issue.

The generally accepted range for THM and HAA recoveries is between 70 and 130 percent. The RPD between expected and reported results for the seven DHBA samples tested was 0.2–16.2 percent (average 7.6 percent), well within this range (appendix D). Errors associated with these analyses include the preparation of 3,5-dihydroxy-benzoic acid (DHBA) standard solution from a solid powder, steps involved with chlorine dosing and quenching, potential loss of the volatile THMs due to bubble formation or air leaks, and variations in the determination of the THM compounds by the laboratory. In addition, because of high concentrations of chlorinated DBPs in the formation potential samples, analyses usually required dilution. Although this additional step could introduce errors, no QA or quality control (QC) issues were apparent in replicate or standard reference samples. Nevertheless, some loss of the more volatile compounds, particularly chloroform, could have occurred during sample transfer for the high-concentration samples requiring this dilution step and may cause a low bias in reported concentrations.

Nutrients—The nutrient data had considerable QA issues, including numerous reported total phosphorus (TP) concentrations that were less than soluble reactive phosphorus (SRP) concentrations and unacceptably high variation (high RPDs) in laboratory replicate "churn-split" samples for SRP, TP, ammonium (NH$_4$), silica (Si), total particulate carbon (TPC), and total particulate nitrogen (TPN) (appendix E). Nitrate concentrations appeared more reliable; laboratory split replicates were within 5 percent of one another. Given the concerns about the quality of the nutrient data, these data were utilized sparingly during analysis, and nutrients were not good explanatory variables during the study.

The average RPDs for TPC and TPN for three laboratory replicate split samples also were relatively high—25 and 34 percent, respectively (appendix E). Given that TOC concentrations were calculated by summing the TPC and DOC concentrations, the potential errors in TPC values could affect

calculations of the percentage of particulate carbon compared to dissolved, which is duly noted.

The appendix data files are included in an Excel© Workbook and are available for download at http://pubs.usgs.gov/sir/2013/0779. This workbook consists of the following worksheets.

Appendix A.   Number and Type of Quality-Assurance Samples.

Appendix B.   Quality-Assurance Data for Dissolved Organic Carbon and Selected Optical Properties.

Appendix C.   Quality-Assurance Data for Disinfection By-Products in Finished Water.

Appendix D.   Quality-Assurance Data for Disinfection By-Product Formation Potentials.

Appendix E.   Quality-Assurance Data For Nutrients and Total Particulate Carbon.

Appendix F.   Spearman Rank Correlations for Select Groups of Samples and Sites.

Appendix G.   Discrete Data Used in the Analysis, Including Watershed, Finished-Water, and Jar-Test Samples.

www.ingramcontent.com/pod-product-compliance
Lightning Source LLC
Chambersburg PA
CBHW081553170526
45166CB00009B/2689